Ai Powered Content Creator

Table of Contents

“I dedicate this book to my wife, Sharmin, and to my only child, Rafan.”

What Is AI? The New Era of AI-Powered Content

The world today is moving rapidly toward a technology-driven future, and at the center of this transformation is the use of Artificial Intelligence, commonly known as AI. Across content creation, media, education, business, healthcare, and digital production, AI is signaling a new kind of future shaped by intelligent technology.

Artificial Intelligence is a technology that enables computers and software to think, make decisions, write, and solve problems in ways that resemble human intelligence. Tasks that once required large, highly skilled teams can now be handled by AI faster, more efficiently, and in many cases with greater accuracy than humans. AI is not limited to content creation alone. Its use is expanding across almost every industry. We have now entered a new era of AI-powered content.

In this era, creating content no longer requires expensive cameras, complex lighting setups, professional software, or large production environments. By using AI, a single content creator can produce cinematic videos, visual designs, voiceovers, animations, scripts, and even complete films if desired. What once seemed impossible for individuals is now becoming part of everyday creative work.

AI-powered content refers to content where artificial intelligence actively assists in the creation process. For example:

- Generating images or videos directly from text
- Automatic video editing and color grading
- AI-generated voiceovers, dubbing, and subtitles

- Smart lighting, camera movement, and visual effects
- Professional-quality output in less time and at lower cost

The greatest advantage of this transformation is the saving of time and cost. Work that previously took hours or required large production teams can now be completed using just a mobile phone or a laptop. As a result, new content creators, freelancers, and students are gaining access to opportunities that were once reserved for large studios and professionals.

Another major benefit of AI-powered content is creative freedom. AI allows human creativity to be expressed on a much larger scale. It helps present ideas more clearly, more precisely, and more powerfully. With AI, content creators can maintain their own stories, ideas, and personal style while producing content that meets modern professional standards.

In the future, the content creation industry will become even more dependent on AI. This is not just a temporary trend. It has evolved into a complete ecosystem for creating content in the modern world.

Many people believe that AI will replace human creativity. In reality, the opposite is true. AI helps enhance human creativity rather than eliminate it. Ideas, stories, and perspectives come from people. AI simply helps turn those ideas into reality faster and more efficiently.
This collaboration between human imagination and artificial intelligence is the true strength of the AI-powered content era.

Followers or Content Quality?

In the digital age, content creation is no longer just a hobby. For many, it has become a profession. Whether it's social media,

YouTube, blogs, or podcasts, one question keeps coming up again and again: is having more followers more important, or is content quality what truly matters?

As a content creator, the most common message or email I receive sounds like this:
"I don't have many followers. Can I still get good views?"
"Is it possible for me to grow an audience?"

My answer is always the same. Everyone starts this way. No one begins with thousands of followers. Every creator starts with one or two people. If the content is good, followers grow naturally over time.

Many people assume that more followers automatically mean success. A large follower base can certainly help increase likes, shares, and views quickly. It can also make a creator more attractive to brands. However, if those followers come without quality content, perhaps through trends, controversy, or short-term virality, that popularity rarely lasts.

On the other hand, content quality creates a strong long-term foundation. Good content solves problems, teaches something valuable, or at the very least provides meaningful entertainment. Quality content builds trust over time. From that trust comes a real audience, not just numbers, but people who actively engage and care.

In reality, when quality is strong, followers come on their own. But having followers alone does not automatically improve quality. That's why focusing on content, not follower count, is the smarter approach. Take time to create content carefully and consistently. Don't worry too much about whether your follower count is low or high. The goal should be to create content people remember, trust, and want to share. In the end, quality is what survives the test of time.

Why Professional Work Is Possible Using Only a Mobile Phone

Many people think, “I can’t afford an expensive camera, so how can I become a good content creator?” This idea is completely wrong. If you look closely, you’ll see that many viral videos are shot using nothing more than a mobile phone.

Even a mid-range smartphone can be enough for content creation. Editing on a mobile phone is often more convenient, and you don’t have to carry extra equipment. With just three things, a mobile phone, a tripod, and a wireless microphone, you can already start creating quality content. I’ve seen many creators producing good work and getting solid views using phones priced between 18,000 and 25,000 taka.

There was a time when professional work meant using a desktop or laptop computer. But times have changed. Modern smartphones are no longer limited to calls, messages, or social media. Thanks to powerful hardware, advanced software, and fast internet, smartphones can now handle many professional tasks with ease.

1. Powerful Hardware

Today’s smartphones come with high-performance processors, large amounts of RAM, fast storage, and advanced GPUs. Because of this, tasks like video editing, graphic design, document editing, and even basic coding are now possible on mobile devices. In many cases, smartphones outperform laptops from just a few years ago.

2. Easy Access to Professional Apps

App stores now offer a wide range of apps built specifically for professional use. Whether it's video editing, photo retouching, presentations, spreadsheets, or digital drawing, reliable apps are readily available. These apps are optimized for mobile use, making them easier and faster to learn.

3. Work From Anywhere

One of the biggest advantages of mobile devices is portability. You can work from home, the office, or while traveling. Without the hassle of opening a laptop, you can reply to emails, edit files, or attend client meetings directly from your phone. This saves time and increases productivity.

4. Cloud and Internet Support

Cloud storage and high-speed mobile internet have made smartphones even more powerful. Saving, sharing, and working on the same project across multiple devices is now simple. Even if work is interrupted, it can be resumed from anywhere without losing progress.

5. Low Cost and Easy Accessibility

With a good smartphone, many professional tasks can be completed without the high expense of buying a separate desktop or laptop. For beginners, a mobile phone is a budget-friendly solution that allows them to start freelancing or content creation without a large upfront investment.

6. Camera and Multimedia Capabilities

Mobile phone cameras have improved significantly. Photography, video content creation, social media marketing, and YouTube video production can all be done using a smartphone. From shooting to editing to uploading, the entire workflow can be completed on a single device.

Because of technological advancements, mobile phones are no longer just communication tools. They have become complete professional devices. While some heavy tasks still require a computer, most daily and many professional tasks can now be handled easily on a mobile phone. With the right skills and apps, building a career using only a smartphone is no longer unrealistic.

Mobile Models, Pricing, and What They Are Best Used For

New content creators often worry about one question above all others:
"How expensive does my phone need to be if I'm not buying a full camera setup?"
They want to know what price range makes sense and what tasks a phone can realistically handle.

The reality is that the market is full of countless phone models, each offering different features. No single phone has everything. That's why choosing the right device can feel confusing. Many people automatically look toward the latest iPhone or Samsung flagship models. However, some apps available on iOS do not exist on Android, and some Android apps either do not run on iPhones or require expensive monthly or yearly subscriptions.

Below is a practical list of popular mobile phones in Bangladesh that work well for content creation. Prices are approximate and based on online stores, official dealers, and local retailers. Prices may vary depending on availability, variants, and market conditions.

Premium / High-End Phones

Best for professional content creation

Samsung Galaxy S25 Ultra

Camera:
Multi-lens camera system with a high-resolution main sensor. Excellent for both photo and video.

Battery:
Large battery with fast charging support. Handles heavy use comfortably for more than a day.

Features:
AI camera tools, 5G support, large and sharp display, extremely powerful performance.

Approximate Price in Bangladesh:
New: BDT 150,000 – 190,000
Used: Around BDT 110,000

Why It's Good:
Top-level performance. Ideal for 4K and 8K video content creation.

Samsung Galaxy S24 Ultra

Camera:
Quad-camera setup with a 200MP main sensor, telephoto, and ultra-wide lenses. Strong low-light and zoom performance.

Battery:
Around 5000 mAh with fast charging support.

Features:
8K video recording, 120Hz refresh rate display, powerful chipset.

Approximate Price:
Official: Around BDT 180,000

Why It's Good:
One of the strongest camera phones available on Android.

Samsung Galaxy S23 Ultra (Used or Refurbished)

Camera:
Professional-grade multi-lens camera with consistent output in almost all lighting conditions.

Battery:
Reliable battery life similar to other Ultra-series phones.

Features:
5G support and flagship-level performance.

Approximate Price:
Used/Refurbished: Around BDT 60,000

Why It's Good:
Flagship power and camera quality at a much lower price.

Apple iPhone Pro Series

(15 Pro / 14 Pro / 13 Pro – New or Used)

Camera:
Excellent video stabilization, Cinema Mode, and reliable video quality.

Battery:
Strong real-world battery life due to iOS optimization.

Features:
Support for professional video apps and workflows.

Approximate Price:
New: BDT 120,000 and above
Used: BDT 80,000 and above

Why It's Good:
Ideal for YouTube, Reels, and short-form video creators. The iOS ecosystem is especially strong for video.

Budget Flagship / Premium Mid-Range

Good balance for photo and video

Google Pixel 10 Pro XL

Camera:
AI-powered multi-lens camera with natural colors and strong detail.

Performance:
Smooth on-device editing performance.

Approximate Price:
Imported or used: Around BDT 90,000

Why It's Good:
Pixel phones remain a top choice for photography-focused creators.

OnePlus 13R (256GB)

Camera:
Strong camera performance in the mid-premium segment.

Battery:
Long-lasting battery with fast charging.

Features:
Good performance for editing apps.

Approximate Price:
Around BDT 60,000

Why It's Good:
Excellent balance between price and performance.

Motorola Moto G Stylus (2025)

Camera:
Decent camera with stylus support for photo and note editing.

Battery:
Around 5000 mAh.

Features:
120Hz display and stylus functionality.

Approximate Price:
BDT 30,000 – 40,000

Why It's Good:
A creative-friendly option on a limited budget.

Budget and Value Options

Suitable for light video and photo content

Samsung Galaxy A36 5G

Camera:
Good daylight photography performance.

Battery:
Reliable battery for everyday use.

Approximate Price:
BDT 30,000 – 40,000

Why It's Good:
A budget-friendly choice for beginners.

Google Pixel 9

Camera:
Strong camera with clean software, good for photo editing.

Approximate Price:
Around BDT 70,000

Huawei Phones (Available in Bangladesh)

(Uses HarmonyOS; Google services may not be available)

Huawei P60 Pro

Camera:
Triple-camera system with strong photo and video quality.

Battery:
Around 4815 mAh with fast charging.

Approximate Price:
New: Around BDT 120,000

Why It's Good:
Excellent for professional photography and zoom performance.

Huawei Mate 80 Series

Camera:
Powerful multi-lens camera setup.

Battery:
5750 – 6000 mAh, suitable for long shooting sessions.

Approximate Price:
BDT 85,000 – 210,000 or higher

Why It's Good:
Ideal for long-form content and heavy usage.

Quick Buying Tips

- **Battery:** 5000 mAh or more makes all-day content work easier
- **Camera:** Multi-lens setup and good low-light performance improve video and reels
- **Performance:** More RAM and a strong chipset ensure smooth editing

All prices mentioned are based on information from Bangladeshi online retailers, official dealers, and local tech websites. Prices may increase or decrease depending on stock availability, variants, memory options, and import conditions.

Traditional Editing vs AI Editing

In today's digital world, content editing is a crucial part of the creative process. Whether it's images, video, audio, or written content, editing plays a major role in shaping the final output. Broadly speaking, editing can be divided into two categories: traditional editing and AI-powered editing.

Traditional editing often takes more time, but it offers a high level of control and refinement. On the other hand, AI editing is advancing rapidly. It saves time and allows creators to use their creativity more efficiently. While traditional editing has long depended on high-performance computers and powerful software, AI tools are now making it possible to continue editing even on mid-range or high-quality smartphones.

In the past, if a computer wasn't powerful enough, certain edits were slow or simply impossible. Today, even with a good mid-range mobile phone, creators can continue editing smoothly and consistently.

What Is Traditional Editing?

Traditional editing is completely manual and human-driven. An editor relies on personal skill, experience, and creative judgment to complete the work.

Examples:

- Manually retouching images in Photoshop

- Cutting and color grading videos in Premiere Pro or Final Cut
- Proofreading or rewriting written content

Advantages:

- Better understanding of emotion and context
- High level of customization
- Complete creative freedom

Disadvantages:

- Time-consuming
- Requires significant time and effort to learn
- Higher cost for large projects

What Is AI Editing?

AI editing uses artificial intelligence to automate or semi-automate the editing process. The software analyzes data and makes decisions on its own.

Examples:

- One-click background removal
- Automatic color correction
- Auto-generated subtitles for videos
- AI-assisted writing correction or content generation

Advantages:

- Extremely fast results
- Beginner-friendly
- Lower cost with good output
- Easy to scale for large volumes of content

Disadvantages:

- Results are not always perfect
- Creative control can be limited
- Human review is often still necessary

Which One Is Better?

There is no single correct answer.

- For films, artistic projects, and premium content, traditional editing is still the best choice
- For social media, fast turnaround work, and bulk content, AI editing is highly effective

The best solution today is a combination of both. Use AI to handle repetitive or basic tasks quickly, then apply traditional editing for the final polish.

Essential Apps, Storage, and Basic Settings

(For mobile-based content creation and editing)

A smartphone is no longer just a communication device. It has become a complete content creation tool. However, to achieve good results, choosing the right apps, managing storage properly, and adjusting basic settings are extremely important.

1. Essential Apps for Content Creation

🎥 Video Editing Apps

- CapCut
- VN Video Editor
- KineMaster

These are highly effective for short videos, reels, and YouTube content.

🖼️ Photo and Graphic Design

- Canva
- Lightroom Mobile
- Snapseed
- Google Nano Banan (Fully AI-powered)
- Seedream 4.5 (Fully AI-powered)
- Kling 01 (Fully AI-powered)
- Reve (Fully AI-powered)
- GPT Image 1.5 (Fully AI-powered)
- Grok Imagine (Fully AI-powered)

These tools are excellent for thumbnails, posters, and color grading.

🎧 **Audio and Voice**

- Dolby On
- Voice Recorder (default or third-party apps)

Useful for voiceovers and clean audio recording.

📝 **Scripts and Ideas**

- Google Keep
- Notion
- Notes (default app)

Ideal for video scripts, content ideas, and planning.

2. Storage Management

(One of the most important aspects for content creators)

Video and photo editing require a lot of storage. Poor storage management can slow down your phone.

Best practices:

- Keep at least 25–30% storage free
- Delete raw footage after editing is complete
- Store exported files in a separate folder
- Turn off auto-download for WhatsApp and Messenger media

Use cloud storage:

- Google Drive
- Google Photos

Back up old projects to keep your phone fast and responsive.

3. Basic Settings Every Content Creator Should Adjust

📷 Camera Settings

- Video resolution: Use the highest supported by your phone
- Frame rate: 30fps or 60fps depending on content type
- Enable grid lines for proper framing

🔋 Battery Settings

- Turn off Battery Saver while editing
- Disable background restrictions for heavy apps
- Keep auto brightness enabled

🔕 Notifications

- Turn on Do Not Disturb while editing
- Limit social media notifications

⚙️ Performance

- Uninstall unnecessary apps
- Restart your phone at least once a week

4. Why These Settings Matter

The right apps, proper storage management, and correct settings lead to:
✓ Smooth editing
✓ Less lag
✓ Professional-quality output

Today, it is completely possible to create professional content for YouTube, Facebook, Instagram, and short-video platforms using only a mobile phone. You just need to know how to use the tools properly.

AI Scriptwriting and Idea Generation

Before using AI for scripting, it's important to understand what AI scriptwriting really is. Content does not become viral simply by speaking randomly in front of a camera. That approach might work once or twice, but if you want consistent views, steady growth, and long-term followers or subscribers, a well-prepared script is essential.

AI scriptwriting is a process where you use prompts to get specific, structured outputs from AI. Content creation includes blogs, videos, social media posts, emails, and presentations. AI scriptwriting helps plan, draft, revise, and refine this process using AI tools such as ChatGPT.

However, before writing a script, you must first decide on the idea. AI can also help with idea generation.

There are several AI tools available for free:

- ChatGPT

- Gemini
- Grok

Personally, I use ChatGPT for all of my work.

How ChatGPT Helps Generate Content Ideas

Simply put, ChatGPT works as an extension of your own thinking. When you provide a theme or topic, AI can:

✓ Generate popular and engaging content ideas
✓ Break ideas into relevant subtopics
✓ Organize concepts into different formats

Don't think of AI as just a text generator. Use it as a creative brainstorming partner.

As a content creator, choose a specific niche or topic. Ask ChatGPT to generate five or ten content ideas related to that niche. One important rule to remember is that beginners should not work across multiple niches at the same time. Focus on one clear topic area, and build from there.

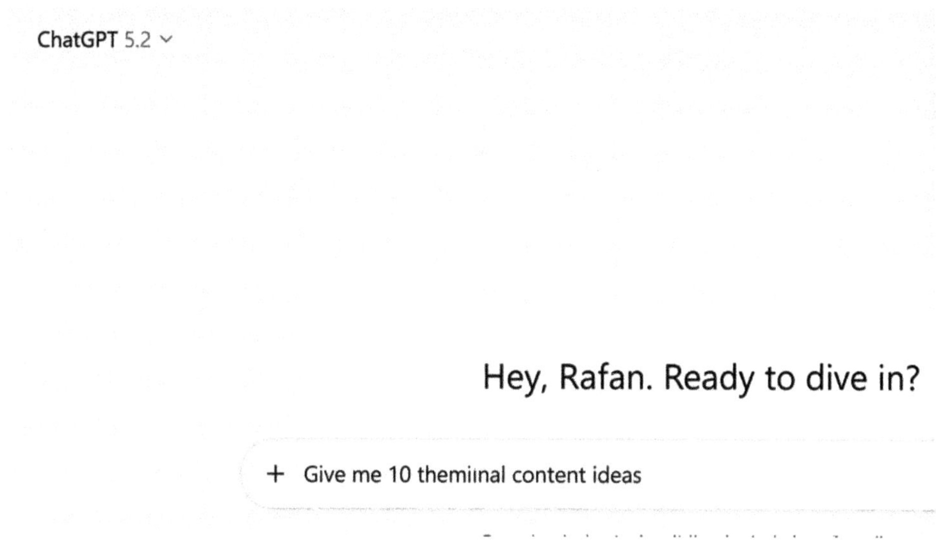

After that, you'll see AI generate content ideas for you in a format similar to the example shown below.

Here are 10 **educational content ideas** that could be useful
making videos, posts, Reels, or easily in short form 👇

1. **1 minute general knowledge facts**
   - One quick "wow" fact, like Earth, science, history.
2. **Why does this happen? (Why series)**
   - Examples: Why is the sea blue, why do ve dream 😁
3. **English Learning Short Tips**
   - Quickly solving a problem pencil ✏️
4. **Science Experiments (In mind or actually)**
   - Simple yet amazing experiment 🧪
6. **A piece of Inspirational Story**
   - Could be short, doesn't need to be true 📜
7. **Mobile Life Hacks**
   - Mobile, internet, online skills and tricks 📱
8. **Beginner Advice (For beginners)**
   - A topic that could be helpful for newbies 🎓
9. **Debunking Common Myths**
   - Testing "Is this true or not?" ✅❎
10. **Q&A/Tips**
    - Answering post or comment questions 💬

↓

“Next, select any one of those ideas and ask AI to turn it into a complete script.”

You’ll then see AI generate the full script for you.

🔎 Quick Tips

Hello friend 👋

Today I'm here to share some quick tips for improving English speaking skills that you can start using right away!

✅ Tip 1: Speak more, don't keep silent

If you stay quiet, you won't learn.

Make mistakes, but don't keep silent.

❌ Happy ≠ ग्रथ (hot this)

✅ I am happy today.

If you make a mistake, someone will correct you.

That's how you'll slowly learn to speak English.

✅ Tip 2: Don't memorize like a parrot

Don't memorize and speak.

You can't speak English by parroting.

Example. **People** ask, what's up? 😊

I'm fine, good, everything is fine.

Every time – use these phrases,

I didn't understand anything.

✅ Tip 3: Use native phrases

Don't use translated English word by word.

Learn native phrases; then English will flow freely.

For instance:

❌ I am **eating rice** now

✅ I am eating now.

❌ I will **call** you after some **time**.

✅ I will **call** you later.

Similarly **rice** = read, call after some time = call later 👍

After that, AI will write the script for you

🔎 Quick Tips

Hello friend 👋

Today I'm here to share some quick tips for improving English speaking sk
you can start using right away!

✅ Tip 1: Speak more, don't keep silent

If you stay quiet, you won't learn.

Make mistakes, but don't keep silent.

❌ Happy ≠ ग्रंथ (hot this)

✅ I am happy today.

If you make a mistake, someone will correct you.

That's how you'll slowly learn to speak English.

✅ Tip 2: Don't memorize like a parrot

Don't memorize and speak.

You can't speak English by parroting.

Example. **People** ask, what's up? 🙂

I'm fine, good, everything is fine.

Every time – use these phrases,

I didn't understand anything.

✅ Tip 3: Use native phrases

Don't use translated English word by word.

Learn native phrases; then English will flow freely.

For instance:

❌ I am eating rice now

✅ I am eating now.

❌ I will call you after some time.

✅ I will call you later

Using that script, you can then create your content. In this way, ChatGPT can be used to create content for any niche or topic.

What Is a Thumbnail? Its Importance and Creating Thumbnails with AI

A thumbnail is the **cover image of a video**
It's the very first thing people see **before they decide whether to click**.

Simply put,
a thumbnail is the poster of your video.

Why Is a Thumbnail So Important?

- People look at the thumbnail first
- If it doesn't catch attention, they skip the video
- A good thumbnail = more clicks

No matter how good your video is, a weak thumbnail can stop people from ever watching it.

Key Features of a Good Thumbnail

1. Big, Clear Text (3–5 words)

Examples:

- *Speak English Fast!*
- *2 Minute English Tips*

Short text works best because it's easy to read even on small screens.

2. Clear Facial Expression

Expressions like:

- Surprise
- Smile
- Thinking or curiosity

Human faces with emotion naturally attract attention.

3. Fewer Colors, Strong Contrast

High contrast works better than too many colors.
For example:

- Yellow or white text
- Dark or simple background

This makes the text stand out clearly.

4. Clean, High-Quality Images

The image must be sharp and clear.
Blurry or low-quality images reduce trust and clicks.

Thumbnail Size (By Platform)

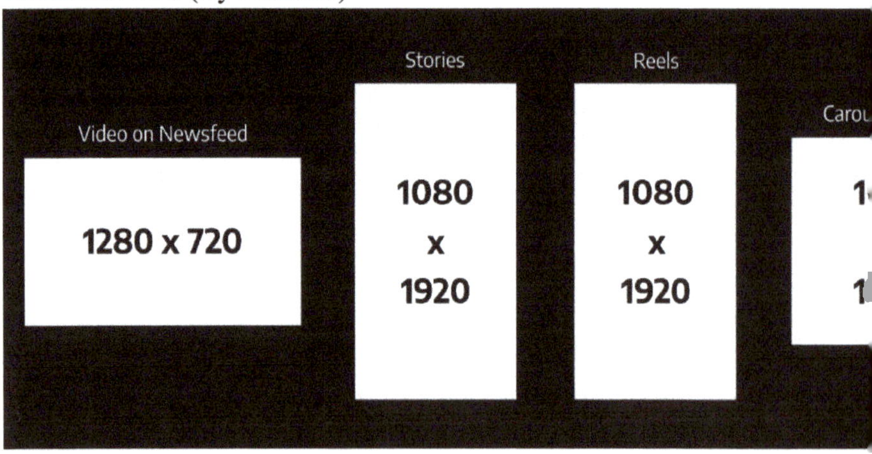

YouTube Thumbnail

- **Size:** 1280 × 720 px
- **Ratio:** 16:9

- **Minimum width:** 640 px
- **Format:** JPG / PNG
- **Max size:** 2 MB
-

  Keep text in the center. Text placed too low may get cropped.

Instagram Reel / Video Cover

- **Size:** 1080 × 1920 px
- **Safe Area (Feed):** 1080 × 1350 px
  Keep important text in the center.

Best Practices (Very Important)

- Use large text with fewer words
- Facial expressions work better
- Avoid too much text
- Always use high-resolution images

If you create one thumbnail in **1080 × 1920**, you can use it on: YouTube Shorts, Facebook Reels, Instagram Reels, and TikTok.

In today's digital world, thumbnails mean views, clicks, and reach.
But do you need to be a graphic designer to create good thumbnails?

The answer is clearly no. From now on, you can create thumbnails using AI.
All you need to know is how to write good prompts.

You'll learn more about prompts in the next chapter.

Creating professional thumbnails with AI, without learning design, is now very easy. Below is the full process explained step by step.

Step 1: Decide the Purpose of the Thumbnail

First, decide why you need the thumbnail:

- YouTube video
- Facebook video
- Shorts or Reels
- Book or course promotion

When the purpose is clear, AI produces better results.

Step 2: Write a Strong Prompt

Prompts are extremely important for guiding AI properly.

Below are both free and paid AI tools for thumbnail creation.

FREE AI Tools: ChatGPT (free and paid)

- Thumbnail ideas
- Text lines

- AI image prompt writing, then generate thumbnails using GPT Image 1.5

Google Nano Banan
Google's AI is very useful for beginners.

- Thumbnail text suggestions
- Design ideas
- Improving AI image prompts

Grok (limited free)

- Bold and eye-catching thumbnail text
- Trend-based captions
- Viral-style wording

Canva (free plan)
The most popular free design tool.

- Ready-made thumbnail templates
- Basic AI design suggestions
- Text and layout setup

Remove.bg (limited free)

- Remove background from faces or products
- Create clean thumbnail cutouts

There are also several paid AI tools such as Midjourney, DALL·E, Kling 01, Leonardo AI, Adobe Firefly, Seedream 4.5, Reve, and others.

Example Prompt for Thumbnail Creation:

Create a high-click YouTube thumbnail, dark background, bold text style, professional look, suitable for an AI tutorial book cover thumbnail.

In your prompt, always mention:

- Background color
- Text style (bold, clean, minimal)
- Mood (professional, energetic, modern)
- Platform (YouTube / Facebook)

Generate the AI Image

After giving the prompt, AI will provide multiple image options. Choose an image that is:

- Clean and clear
- Simple but eye-catching
- Focused on a single subject

Keep the Text Simple

Too much text reduces clicks. The best approach is:

- 2–4 words
- Large, clear font
- High-contrast colors

Examples:

- AI-Made Thumbnails
- No Design Skills Needed
- Click-Ready Thumbnails

Refine Using AI

The first image is not the final one. Ask AI again to:

- Make the text clearer
- Increase face focus
- Simplify the background

This makes the thumbnail look more professional.

Set the Correct Size and Format

Adjust the size based on the platform:

- **YouTube:** 1280 × 720
- **Facebook:** 1200 × 675
- **Book promo thumbnail:** Square or custom size

Step 7: Final Check

Before using it for posting or printing, check:

- Is it readable on mobile?
- Is any text getting cut off?

- Does it stand out at first glance?

If it's clear at a single glance, the thumbnail is successful.

For thumbnails, always decide the design based on aspect ratio first.
Check all commonly used aspect ratios below before designing.

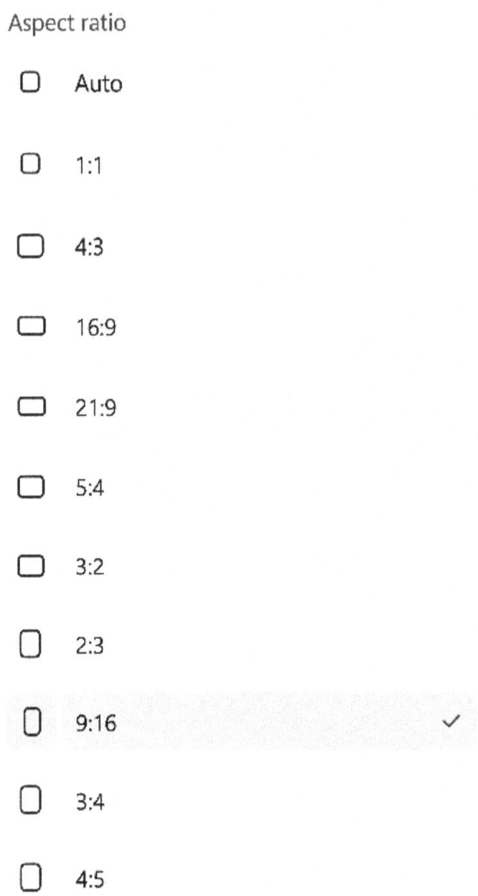

Since most people now focus on creating Reels, thumbnails should almost always be designed in the **9:16 ratio**.

What Is a Prompt, How to Structure It Properly, and How to Use It

A prompt is a written instruction, question, or request given to an artificial intelligence system to tell it what task to perform. Simply put, a prompt explains what the user wants the AI to do and how they want it done.

AI systems analyze the prompt and then provide information, generate text, translate content, perform analysis, or complete decision-based tasks. The clearer and more well-structured the prompt is, the more accurate and effective the AI's response will be.

Structure of a Good Prompt

An effective prompt is usually made up of several key elements.

1. Define the Objective

First, clearly state what result you want. For example: writing, analysis, explanation, a list, or a summary.

2. Context

If needed, provide background or relevant information so the AI can understand the topic properly.

3. Specific Instructions

Mention the length, language, tone, level (simple, academic, professional), or structure you want.

4. Constraints or Conditions

Clearly stating what should be included, what should be avoided, and any rules to follow helps improve the output.

Example

Weak prompt:
"Write about climate change."

Good prompt:
"Explain the causes and effects of climate change in 100–150 words, using simple language, for higher secondary students."

Uses of Prompts

Prompts are highly versatile and important across many fields.

1. In Education

Prompts are used to explain lessons, create notes, prepare question answers, and develop initial research ideas.

2. Writing and Creative Work

Prompts play a key role in creating stories, poems, essays, scripts, and content.

3. Professional Work

Prompts are used for writing emails, creating reports, preparing presentations, and analyzing data.

4. Technology and Programming

Prompts are useful for writing code, finding errors, explaining algorithms, and creating software documentation.

Prompts are the foundation of using AI effectively. A well-structured prompt helps unlock the full potential of AI. To get accurate and high-quality results, developing strong prompt-writing skills is essential.

If you're still unsure where prompts are used, here's a simpler explanation:
Whenever you ask ChatGPT to do something, the instruction you give is called a prompt.

An example is given below.

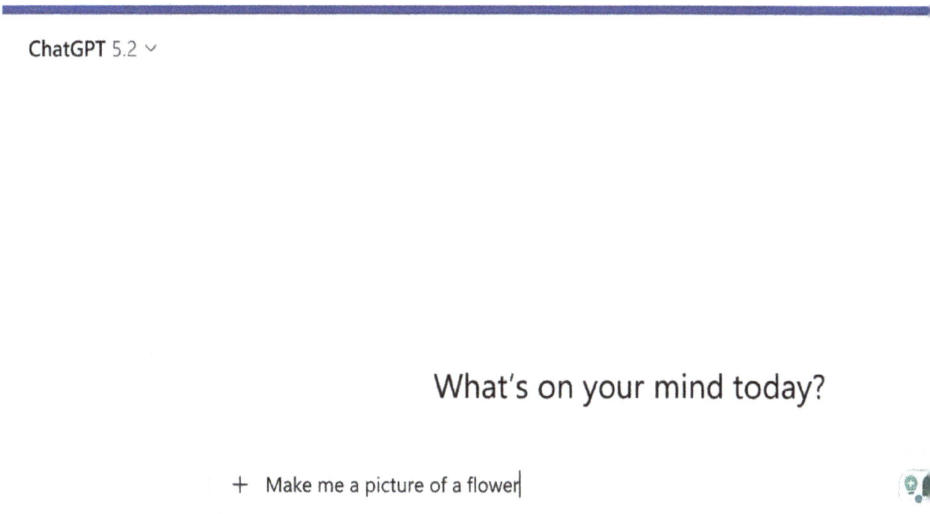

"The instruction you gave here is called a prompt. Once AI receives your prompt, it starts working for you and can produce images, videos, music, or other content.

Based on your prompt, AI created an image for you. This is how AI works.

Why Is YouTube / Facebook Thumbnail Design Important?

The reason YouTube or Facebook thumbnail design is important is actually very simple.

1. First impression happens here

A thumbnail is the face of your content. It's the first thing people see before opening a video or post. If it's not attractive or eye-catching, they will simply scroll past.

2. The biggest tool for increasing clicks

The same content performs much better with a good thumbnail. A strong thumbnail increases the click-through rate, meaning more people actually click to watch. When that happens, the algorithm also shows the content to more users.

3. Standing out in the crowd

YouTube and Facebook are extremely crowded platforms now. There are hundreds of videos and posts everywhere. Without a strong thumbnail, your content gets lost in the crowd.

4. Communicating the message at a glance

A good thumbnail explains what the content is about in just one second. People instantly think, "This is something I need to watch."

5. Builds brand value

When you use a consistent style, color, and font, people start recognizing your content just by looking at the thumbnail. Over time, this builds trust and a professional image.

6. Increases watch time and engagement

When people click with genuine interest, they are more likely to watch the video till the end. Likes, comments, and shares also increase.

If you visit my Facebook page, you'll notice that the videos look more attractive because of the thumbnails. Below, I've shared some thumbnail examples from my page to help you understand.

Which AI Is Best for Mobile Thumbnail Editing?

Without going into too much detail, you can use **Google Gemini Nano Banana** to create thumbnails. It can be used for free, and once the limit is reached, you can simply log in with another email to get a new limit.

To use Google Gemini Nano Banana, scan the QR code below. You can also find the **Gemini app** on your mobile phone in the App Store or Play Store, for both iPhone and Android.

After opening Google Gemini Nano Banana, tap on 'Create Image' as shown in the image below, then write your prompt

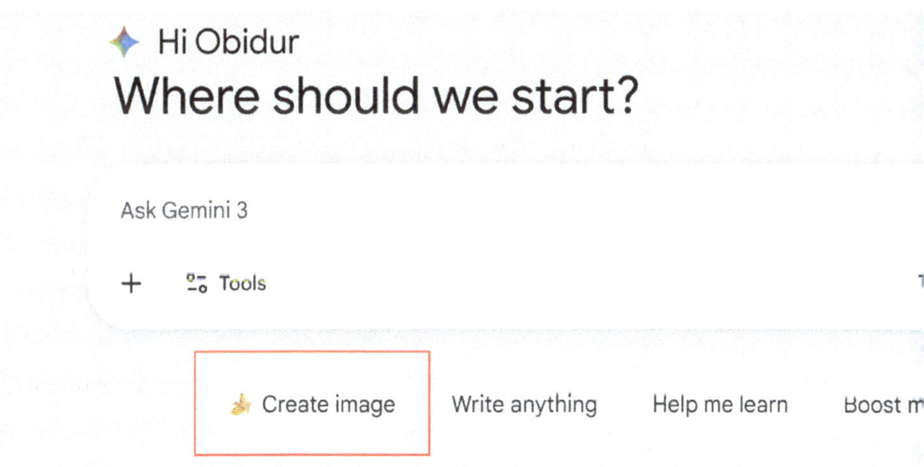

Now write your prompt. First, tap the plus (+) icon below and upload a photo of yourself or the image you want to use as the main background of the thumbnail. Then give your instructions similar to the prompt shown below

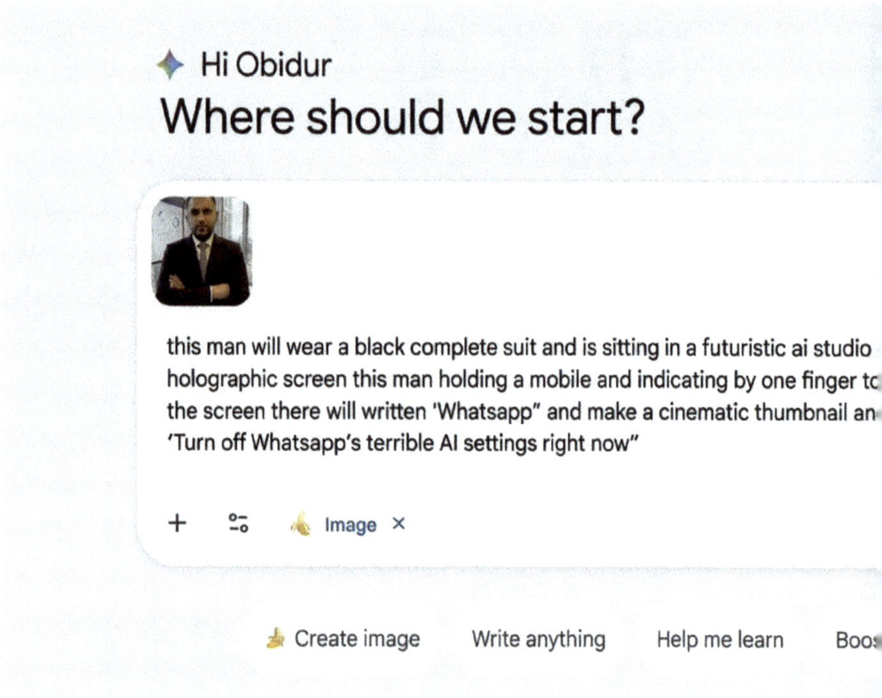

AI will then start creating the thumbnail for you for free, and in less than one minute, it will generate a clean and attractive thumbnail at no cost

In this case, if the reference image you upload has a **9:16 aspect ratio**, the thumbnail output will also be **9:16**. If the reference image is **16:9**, the thumbnail output will follow the same ratio.

AI-Powered Image Background Removal, Upscaling, and Image Expansion

In digital content creation, removing an image background has become a very common but important task. In the past, this required learning complex software, spending a lot of time, and having technical skills. Now, AI-powered background removal technology can complete this task in just a few seconds.

AI technology automatically detects the subject inside an image and separates it from the background. This saves time, simplifies the process, and produces a much cleaner and more professional result. First, AI identifies the person, product, or main object in the image.

Then it detects the edges or outline of the object and separates the background from the subject, cutting even fine details like hair, hands, or product corners cleanly and accurately.

This entire process happens automatically, without requiring the user to do anything manually.

For this type of work, you can use AI tools such as **Canva, Adobe, Grok, Gemini, and Google Nano Banana**.

To remove a background using AI, I would again recommend using **Google Nano Banana**. After opening Nano Banana, first upload the image whose background you want to change. Then clearly mention what kind of background you want to replace it with. I'm giving an example myself so you can understand easily.

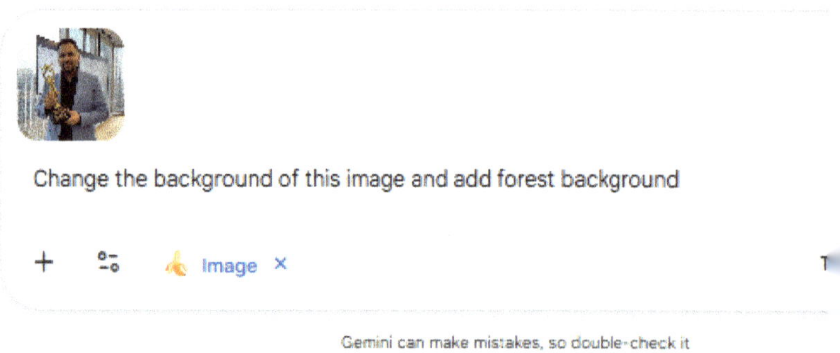

As you can see below, AI has adjusted the skin tone properly, matched it with the background, and completed the task for you in just 30 seconds, completely free.

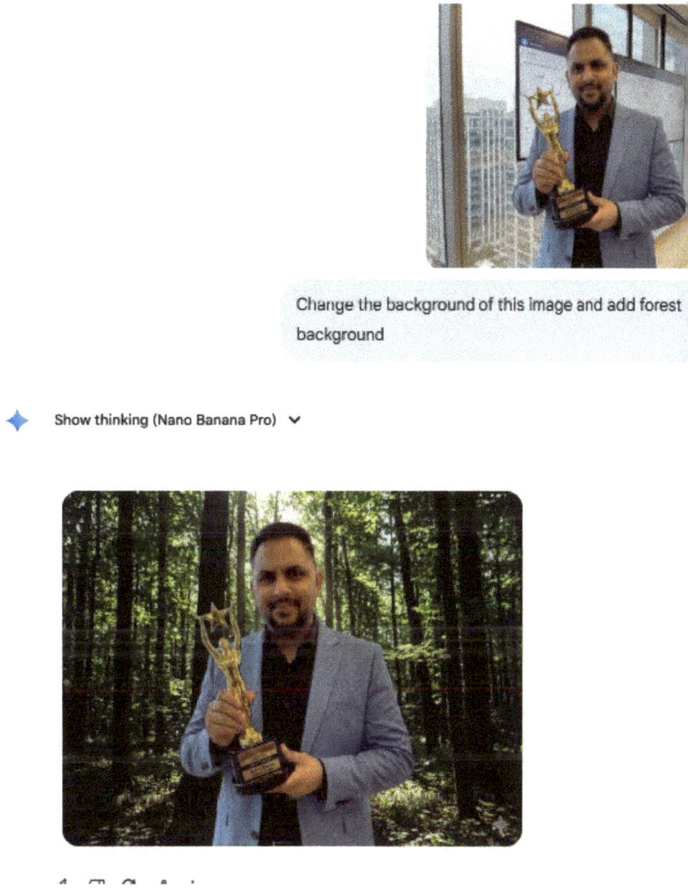

What Is Image Upscaling?

Low Resolution (Bad)

High Resolution (Good)

What Is Image Upscaling?

Image upscaling means enlarging a small or low-resolution image while trying to maintain the best possible quality. In simple terms, it means increasing the number of pixels in an image.

A digital image is made up of countless tiny dots called pixels. The higher the resolution, the more pixels an image has, and the clearer it looks. When upscaling an image, the size increases, but the main challenge is creating new pixels in a way that doesn't make the image look blurry or broken.

Why Is Image Upscaling Needed?

Image upscaling is used for many reasons, such as:

- Preparing old or low-quality images for printing
- Keeping images clear when displayed on large screens

- Achieving better visual quality in movies, games, or YouTube videos
- Seeing more detail in medical imaging or satellite images

How Image Upscaling Works

1. Traditional Upscaling

This is the oldest method. The software looks at nearby pixels and creates new ones based on estimation. Common methods include:

- Nearest Neighbor
- Bilinear
- Bicubic

These methods work quickly, but the downside is that images often become blurry or lose detail when enlarged. That's because no new information is actually created, pixels are just guessed and filled in.

2. AI-Based Upscaling

The most popular modern method uses artificial intelligence. This approach works differently.

AI models are trained on millions of high-quality images. They learn:

- How eyes, hair, text, and edges should look

- What kind of details belong in different areas

When a low-resolution image is provided, AI doesn't just stretch the pixels. It intelligently recreates missing details. As a result, the image looks sharper, cleaner, and more realistic.

Steps in AI Upscaling

1. Analyze the input image
2. Detect objects, edges, and textures
3. Predict possible missing details
4. Generate a new high-resolution image

This entire process happens within seconds, powered by complex algorithms and neural networks.

Limitations of Image Upscaling

- Extremely poor-quality images cannot be fully restored
- AI may sometimes generate details that don't perfectly match reality
- Excessive upscaling can make images look unnatural

The Future of Image Upscaling

Image upscaling technology continues to improve. In the future:

- Old historical photos will look more realistic

- Real-time video upscaling will become common
- Automatic super-resolution will be built directly into mobile cameras

Image upscaling is not just about making images bigger. It is an intelligent process of creating higher-quality visuals from limited information. As technology advances, image upscaling will play an even more important role in the digital world.

To upscale images using AI for free, scan the QR code below.

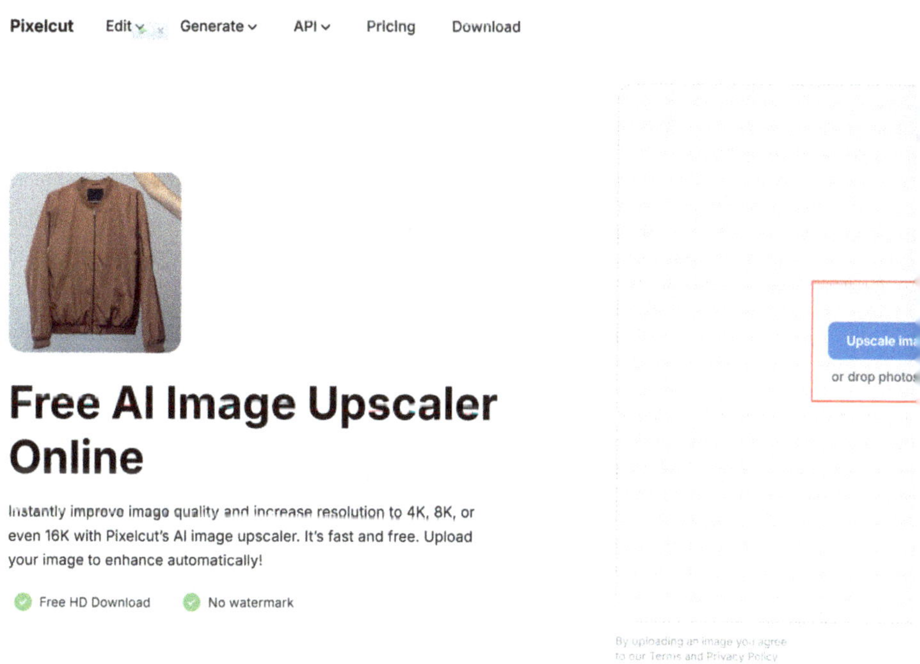

Once the image is uploaded, you'll see options to upscale it to 2× or 4×.

After the image is upscaled, you'll find the Download option at the top right. Download it from there."

What Is Image Expand?

Image expand means adding new space around an image. The main subject remains unchanged, but new background or elements are generated outside the original image. This is often called outpainting or generative expand.

Simply put:
Making an image bigger = Upscale
Creating new areas outside the image or resizing the canvas = Expand

Difference between Upscale ar

Subject	Upscale	
Function	• Increases image quality	• Ir
Pixels	• Increases inside	• A
Original Image	• Remains the same	• R
New Content	• No	• Y

Why Is Image Expand Needed?

Image expand is used when:

- The image frame becomes too small
- A different size is required for social media or posters
- Extra space is needed for video or design work
- A cropped image needs to look natural again

Example:
You have a close-up photo of a person. To create a poster, you need empty space around the subject. Image expand creates that extra space in a realistic way.

How Image Expand Works

1. Understanding the Image Context

AI first analyzes what's inside the image:

- People, sky, trees, walls, lighting
- Colors, textures, and direction

2. Predicting Possible Surroundings

Based on what exists inside the image, AI predicts what could exist outside it. For example:

- If there is sky → it continues the sky
- If it's an indoor scene → walls, windows, or lighting

3. Creating New Areas

AI then gradually generates new pixels so the image looks like it was originally larger.

Traditional Expand vs AI Expand

Traditional Methods

- Adds empty space or blur
- Uses copy-paste stretching
- Results look unnatural

AI-Based Expand

- Creates realistic backgrounds

- Matches lighting and shadows
- The change is not easily noticeable

Limitations of Image Expand

- With very little visual information, AI may make incorrect assumptions
- Complex objects can become distorted
- The new areas are generated based on prediction, not real data

AI Voice Over Creation: Cloning Your Own Voice from Text

AI Voice Over means using artificial intelligence to create natural, human-like speech. Here, no person needs to speak into a microphone. Instead, the computer reads written text and converts it into voice.

Many content creators do not want to use their own voice, and some are unable to do so. Others struggle with hesitation or lack confidence while speaking. In such cases, AI voice over becomes extremely useful.

Simply put:
Text → AI → human-like voice

This voice can be male, female, child, elderly, emotional, or neutral in tone.

Why Is AI Voice Over Used?

AI voice over has become very popular because:

- Videos, reels, and YouTube content can be created quickly

- No studio, microphone, or voice artist is required
- The same script can be produced in multiple languages and voices
- Both time and cost are reduced

Used in:

- YouTube videos
- Advertisements
- Audiobooks
- Online courses
- App and game instructions

How Does AI Voice Over Work?

1. Text Input

First, the user provides written text. This text is later converted into voice.

2. Text-to-Speech (TTS) Process

AI analyzes words, sentence structure, pauses, and pronunciation.

3. Voice Model Selection

AI uses pre-trained voice data. Here, you can choose:

- Voice type
- Speaking speed
- Emotion (happy, serious, calm)

4. Audio Generation

Finally, AI generates a digital audio file that sounds like a human voice.

What Is AI Voice Cloning?

Voice cloning is a more advanced form of AI voice over. In this process:

- A sample of a person's voice is provided
- AI learns that voice

- New text is then spoken in the same voice

Used for:

- Dubbing
- Brand voice
- Audiobook series

Ethics and permission are very important in this case.

Limitations and Cautions

- Not all AI voices sound fully human yet
- Mispronunciations can occur
- Cloning a voice without permission is unethical and risky

The Future of AI Voice Over

In the future, AI voice over will be:

- More emotionally expressive
- Capable of real-time conversations
- Used for live voice generation in movies and games

Traditional Voice Over vs AI Voice Over

Subject	Traditional	AI
Time	More	Very Less
Cost	More	Less
Studio	Needed	Not Needed
Changes	Hassle	Easy
Scale	Limited	Huge

You can create voice overs for free using text only, with male, female, boy, or girl voices through Google Voice Over. However, for voice cloning, you will need to use a paid service.

To use free voice over, scan the QR code below.

After logging in with your email, you will see an interface like the one shown below. Please look at the image below.

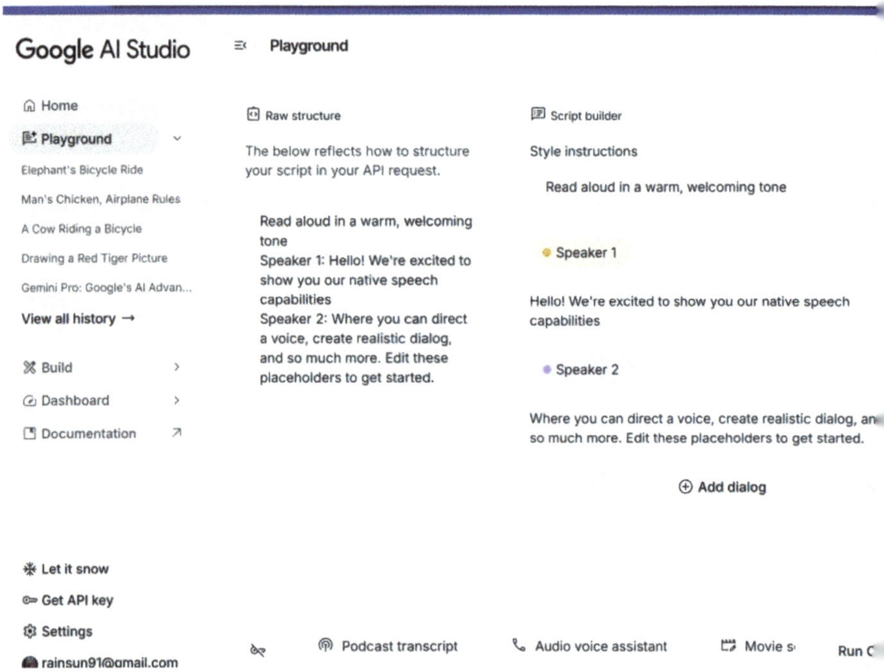

If you want to create a voice using a single-voice prompt, click on 'Single Speaker Audio' as shown in the image. Then, from the speaker settings marked in the red box, select the type of voice you want.

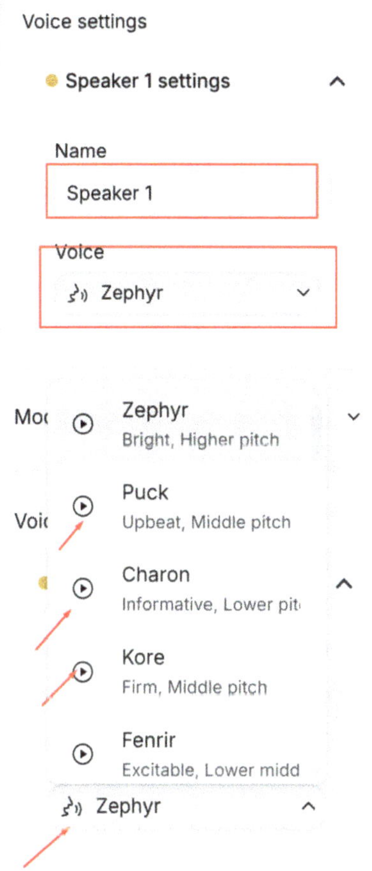

Next, enter the text in the red box shown in the image, which AI will convert into voice. After entering the text prompt, click 'Run', and the voice will be ready.

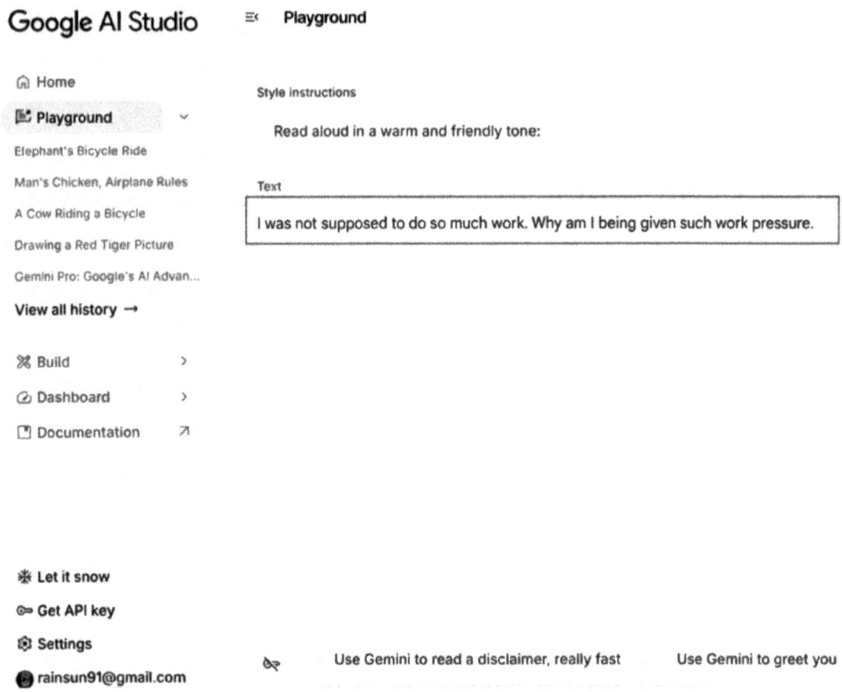

After that, you'll see an option to play the voice as shown in the image below. Clicking the three dots will give you the download option.

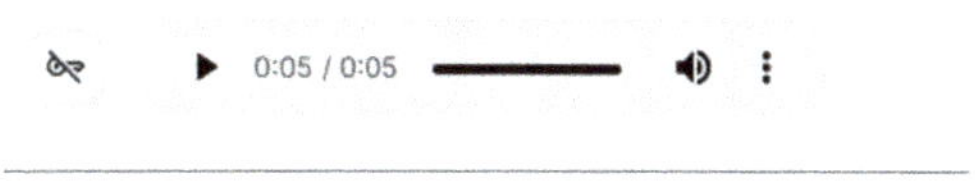

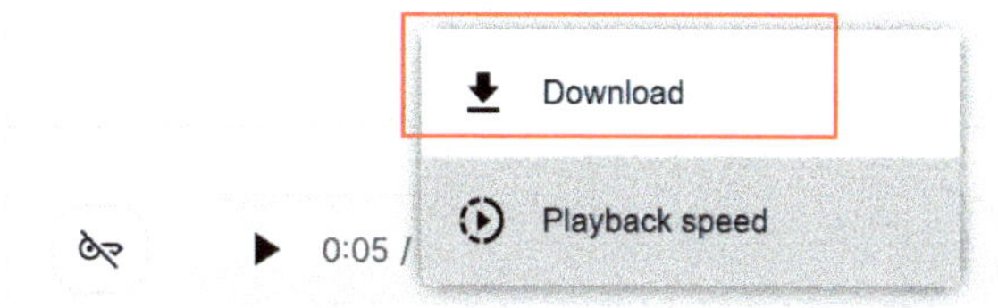

If you want to create two voices using text, first select 'Multi Speaker Audio'. Then choose the type of voice you want, such as male or female, and enter the voice text accordingly

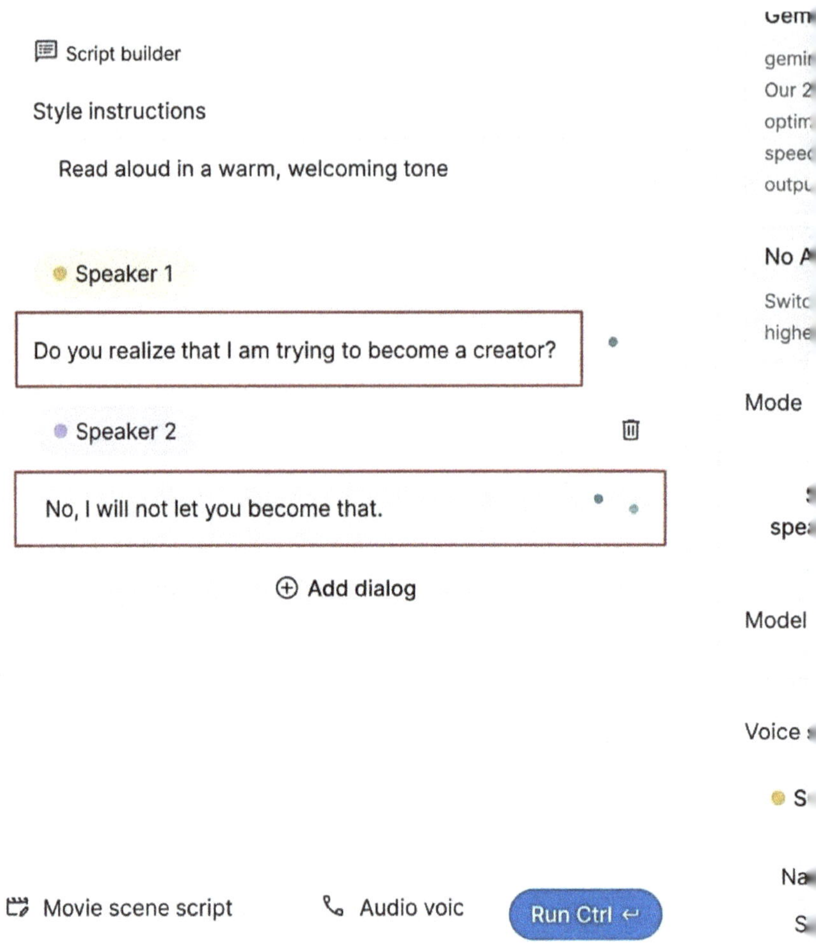

Then run it using the same method shown earlier and download the output.

What Is AI-Powered Voice Cloning?

AI-Powered Voice Cloning is a technology that uses artificial intelligence to accurately replicate a specific person's voice. It doesn't just copy speech, it also imitates tone, pronunciation, speaking speed, and even emotions.

Simply put:
Human voice → AI learns → new text spoken in the same voice

Why Is AI Voice Cloning Important?

This technology is important because it:

- Turns a voice into a digital asset

- Allows new content to be created without repeated recording
- Breaks language and time barriers

Used in:

- Audiobooks
- YouTube and documentaries
- Games and animation
- Advertising and brand voice
- Language translation and dubbing

How AI-Powered Voice Cloning Works

1. Voice Sample Collection

First, audio samples are collected from the person whose voice will be cloned. Usually required:

- Clean recordings
- Different sentences
- Various tones and speaking speeds

2. Voice Analysis

AI analyzes the characteristics of the voice:

- Pitch
- Tone

- Pronunciation style
- Speaking rhythm

At this stage, AI creates a “voice signature.”

3. Model Training

An AI model is trained using this data.
The model learns:

- How the person speaks
- Where emphasis is placed on words

4. Text-to-Voice Conversion

When new text is provided, AI converts it into audio using the cloned voice.

Emotion-Aware Voice Cloning

Modern AI doesn’t just copy voices, it understands emotions as well, such as:

- Happiness
- Sadness
- Anger
- Calm tone

This makes the voice sound more natural.

Risks and Ethical Considerations

AI voice cloning is very powerful, which also brings risks:

- Using a voice without permission can be illegal
- It can be misused for fake calls or fraud
- Trust and credibility can be damaged

Responsible use is essential:

- Clear consent
- Ethical guidelines
- Transparency

Future Possibilities

In the future, AI voice cloning may:

- Support live conversations
- Speak multiple languages using one voice
- Act as a personal digital voice assistant

For voice cloning, I recommend using **ElevenLabs**. There are many AI tools available online, but this one performs the best among them. Scan the QR code below to visit the website using your mobile or computer browser, then click the red box to sign up using your Gmail account.

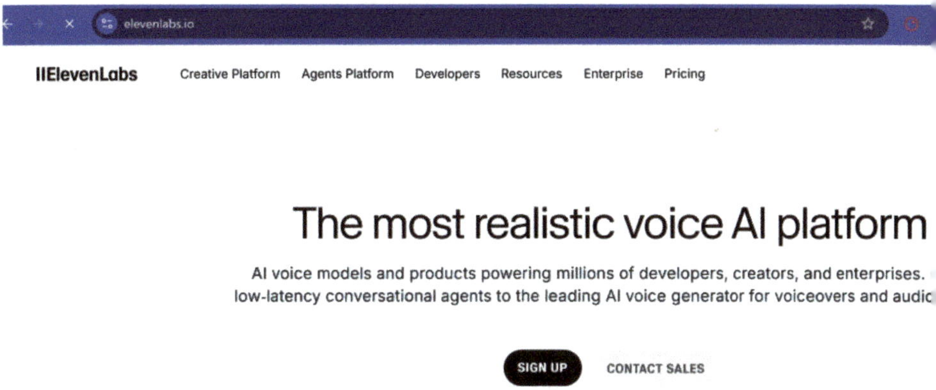

After signing up and logging in, select the Creative Platform.

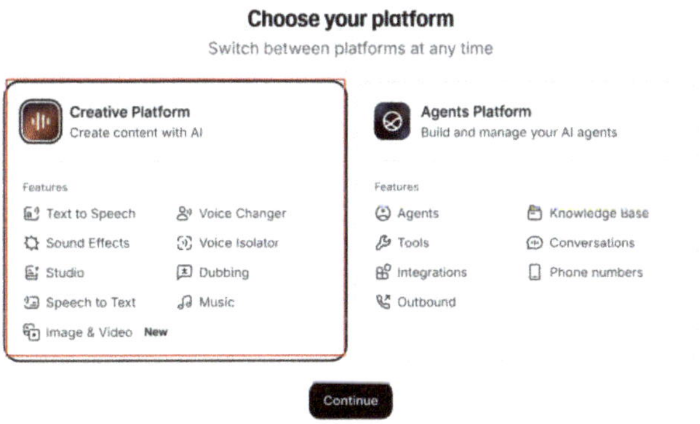

Then simply select the remaining options according to your preference. There won't be any issues

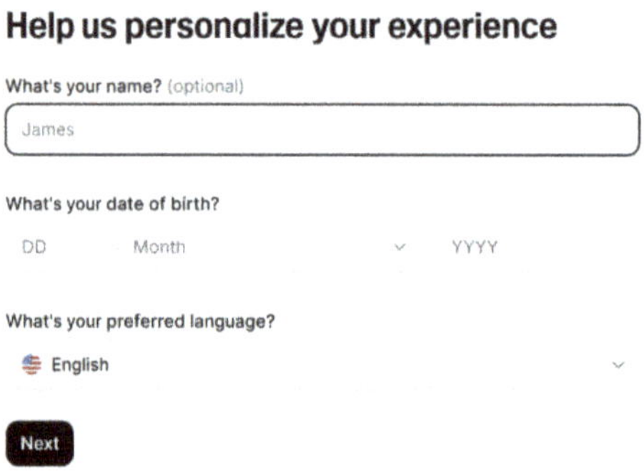

Fill in your personal details properly, and then you'll see an interface like the one shown below and receive 10,000 credits. Since this is for professional use, you can either purchase a subscription or continue using it by changing email accounts.

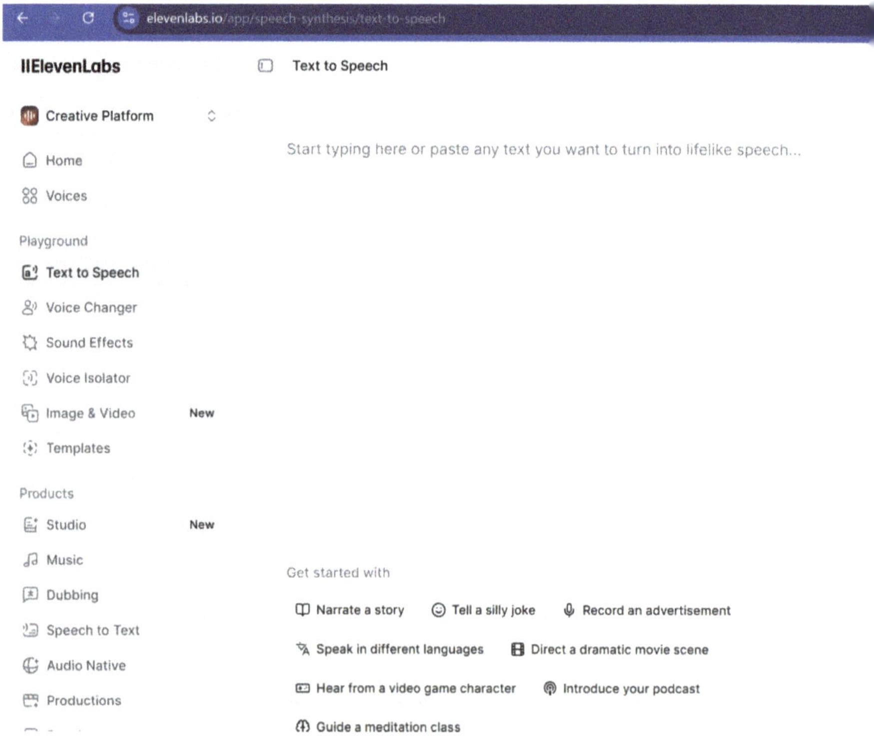

"Next, click on 'Voices' on the left side as shown in the image below, and you'll find the voice cloning option on the right

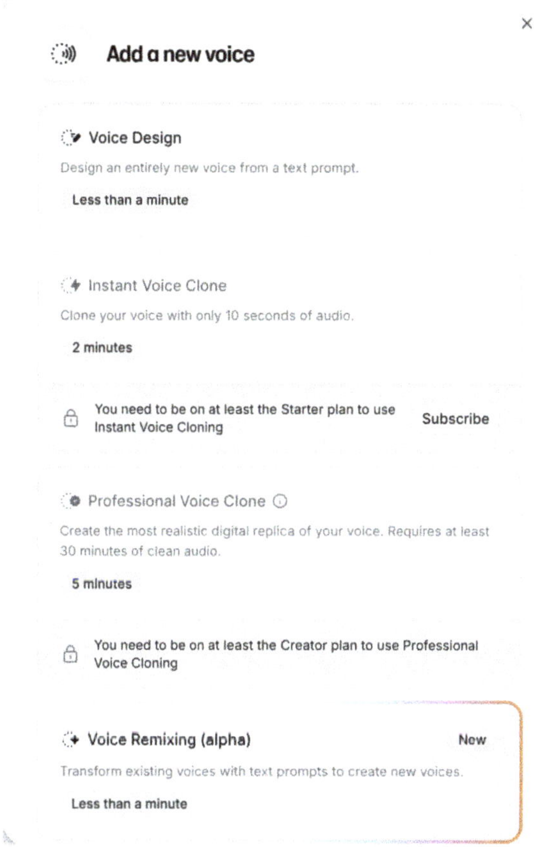

After that, you'll see several options.

1. Voice Design

Voice Design is a method where you can create a completely new voice without using any real human voice.

Here, the user defines the voice characteristics through text, such as:

- Soft or deep voice
- Male, female, or neutral
- Calm, confident, or storytelling tone

Based on these instructions, AI creates an entirely new artificial voice.
This process is very fast and usually takes less than a minute.

It is used when you don't need your own or a familiar person's voice, but instead want a generic or brand voice.

2. Instant Voice Clone

Instant Voice Clone allows you to copy someone's voice using a very short audio sample.

In this case:

- Only about 10 seconds of voice audio is needed
- AI learns the structure, pronunciation, and tone of the voice
- New text can then be spoken in that voice

This method works quickly, but the accuracy of the voice is medium level.
Using this feature usually requires at least a basic paid plan.

It is suitable for quick tasks, demos, or temporary content.

3. Professional Voice Clone

Professional Voice Clone is the most advanced and realistic voice cloning method.

In this process:

- At least 30 minutes of clean, noise-free audio is required
- AI deeply analyzes voice characteristics
- It replicates speaking style, emotion, and natural pauses

The resulting voice sounds very close to the original human voice.
Training the AI model takes some time, but the output quality is extremely high.

This method is used for:

- Audiobooks
- Brand voices
- Dubbing
- Long-term content creation

4. Voice Remixing (Alpha)

Voice Remixing is an experimental (Alpha) technology.

In this method:

- An existing voice is taken
- It is modified through text instructions
- A new variation of the voice is created

For example:

- Making the same voice sound younger
- Making the tone warmer or deeper

This feature is mainly used for creative work and experimentation.

AI-powered voice creation technology allows human voices to be digitally created, cloned, and transformed. When used properly, it makes content creation easier, faster, and more efficient. However, ethics, consent, and responsible use are always important.

If you want to use **instant voice cloning**, you must purchase a subscription from them, which costs around **$5**. If you don't have a card, you can also purchase it through https://subscriptionsmartbd.com/ using bKash.

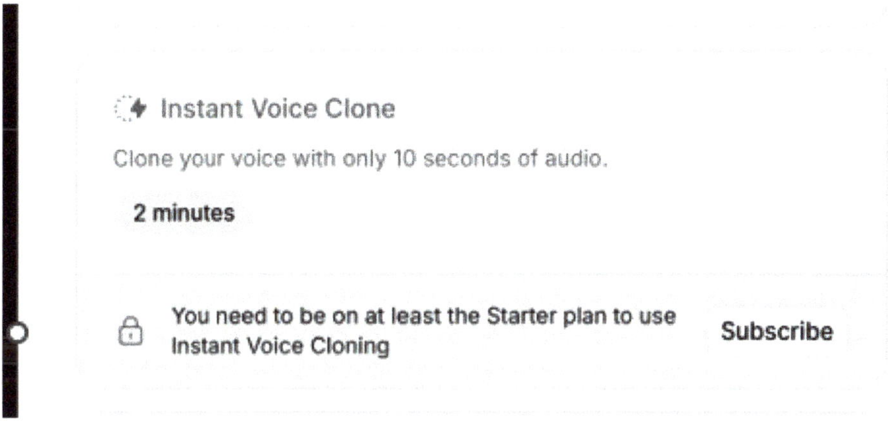

After that, you'll see an interface like this

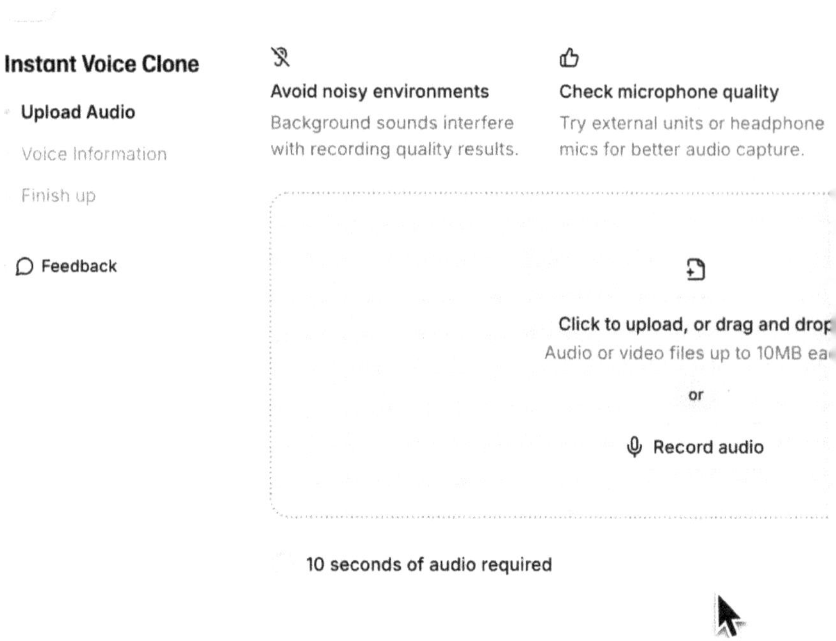

There, you can either record audio or upload a 10-second file that contains your voice. I'm clicking on 'Record Voice' and recording directly using my mobile or computer microphone. Click the Start button and record for 10 seconds.

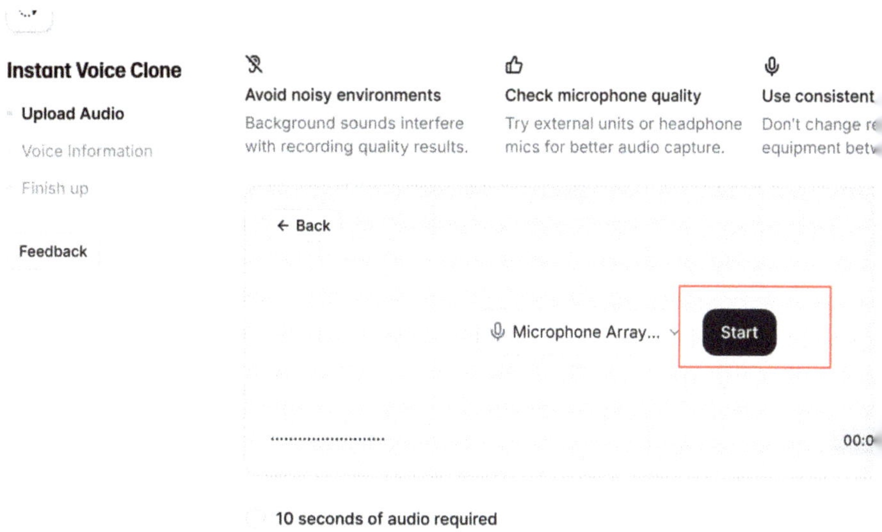

After the recording is complete, click the settings option as shown on the screen below, then click 'Next

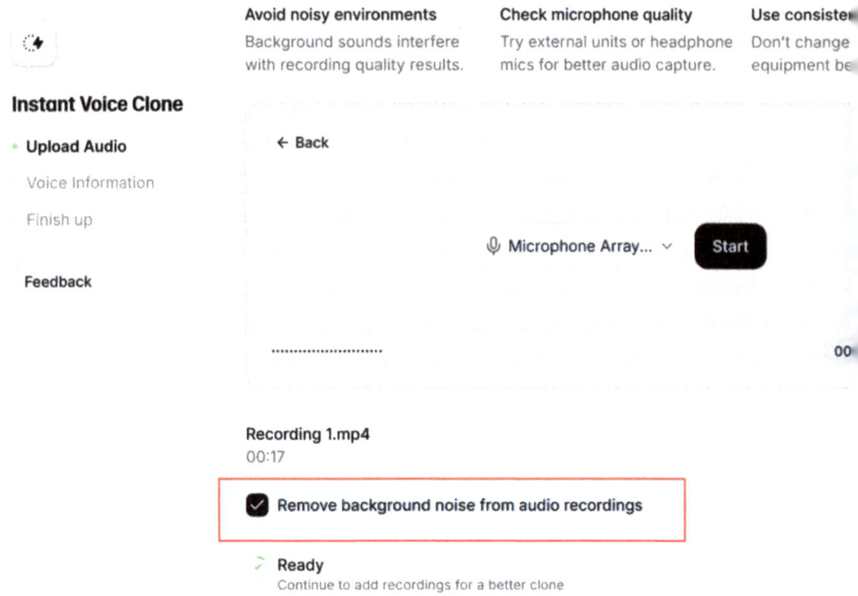

Next, fill in your details as shown in the information below and click 'Save Voice

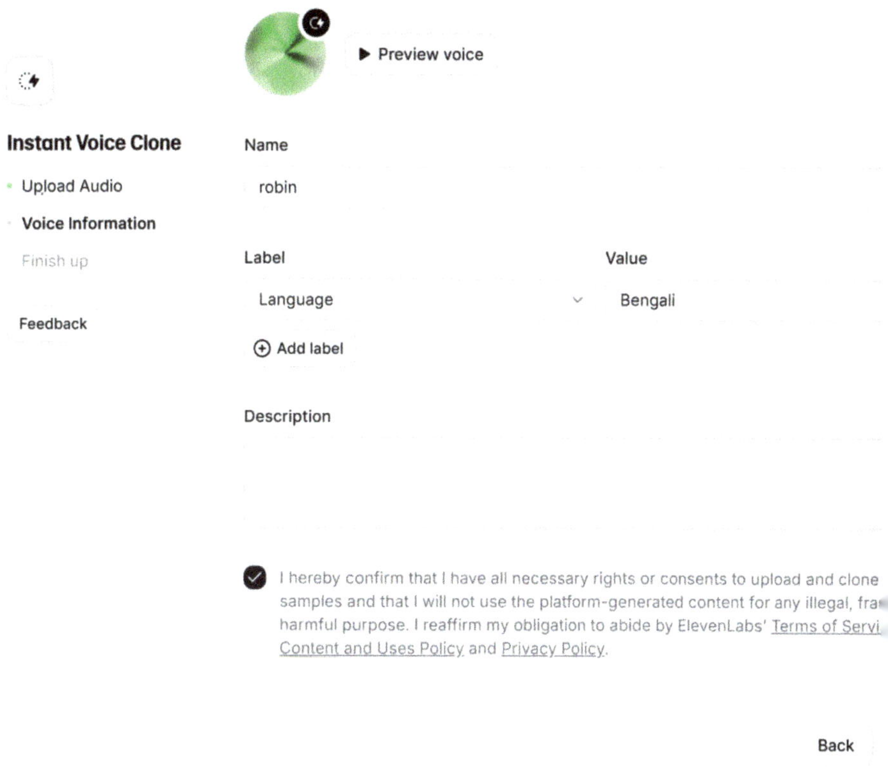

Now you'll see that, just like in the image below, you can create audio in your own exact voice simply by typing text. Click

'Generate Speech', enter your text, and the audio will be created

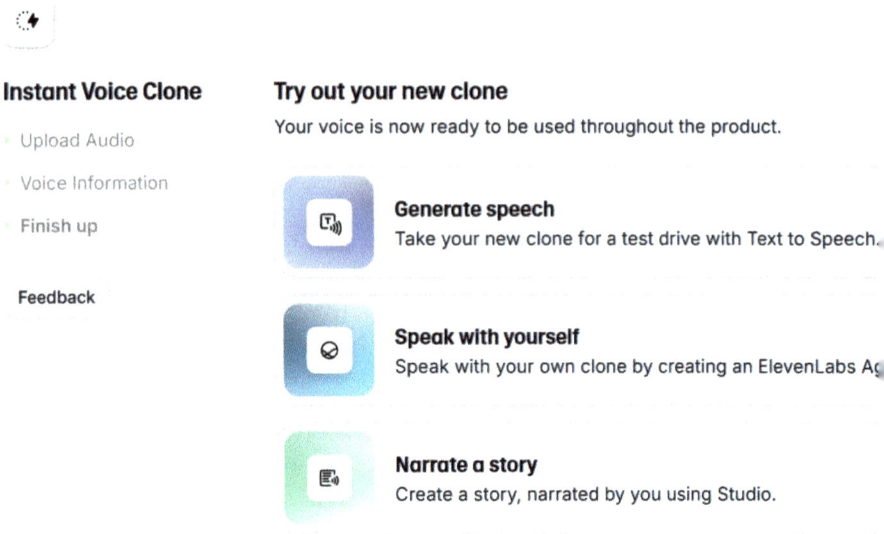

Now, as shown, write the English text prompt in the first red box, select 'Robin' in the other red box, and then click 'Generate Speech' in the black bo

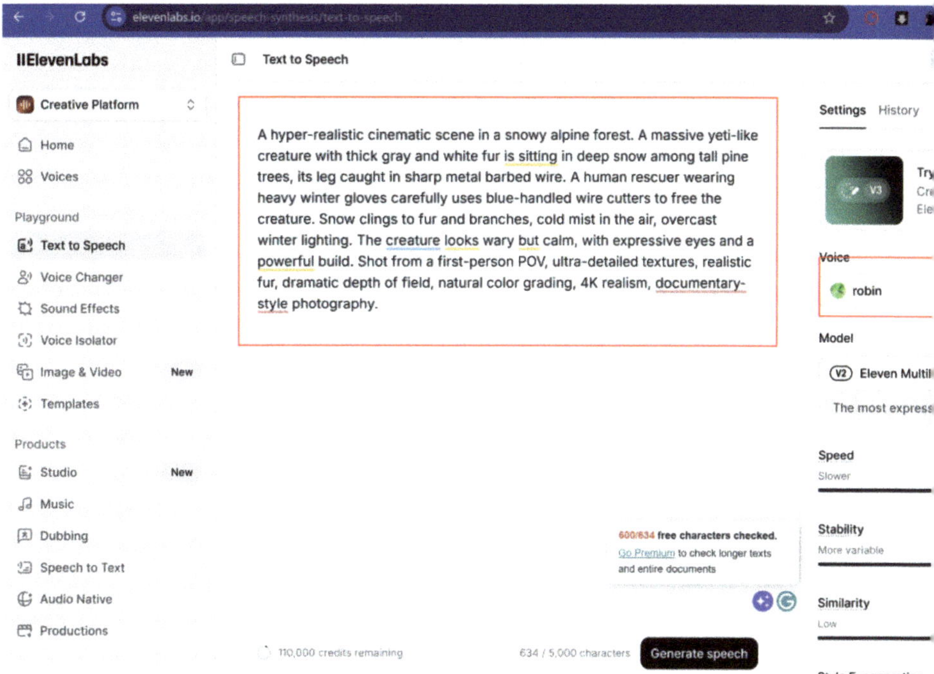

Then you'll see that your speech has been generated. Click the area marked in red below, and it will be downloaded.

However, if you want the voice to speak in Bangla or generate Bangla speech, you need to click on the 'V3 Alpha' option marked below.

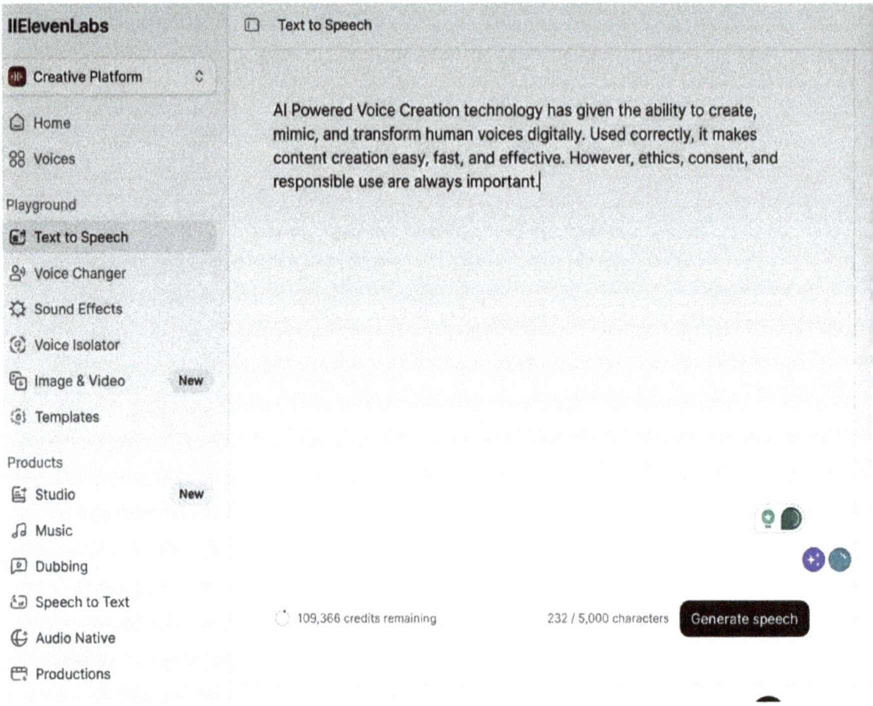

After that, you'll see an interface like this. Clicking 'Generate Speech' will make AI start speaking in your voice. It will provide two output versions of your voice. Click the red-marked option to download whichever one you prefer.

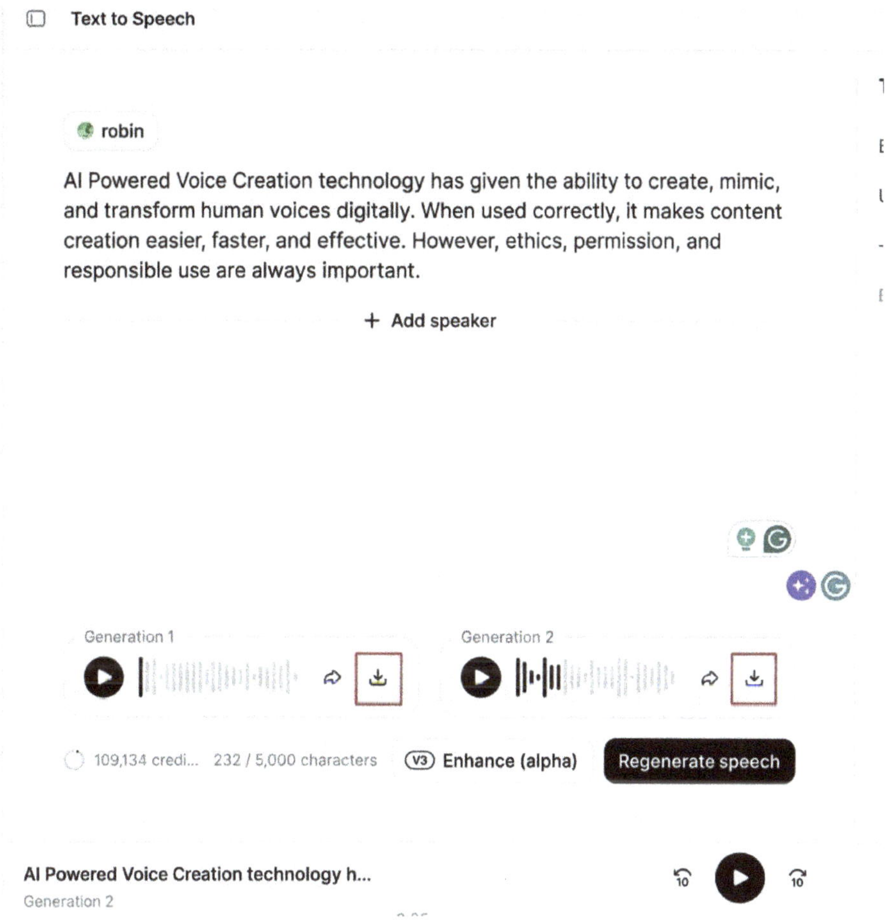

To purchase subscriptions for this type of AI service using Bangladeshi currency, you can visit www.subscriptionsmartbd.com. It is a third-party subscription purchasing site. I use it myself, so I'm sharing it with you as well.

Mobile Video Editing

In today's digital age, video content is no longer just for entertainment. It plays an important role in education, marketing, journalism, and personal branding. In the past, video editing meant powerful computers and complex software. Now, smartphones have become complete video editing studios.

Earlier, creators had to carry cameras and multiple lenses. Today, much of that work can be done using the small mobile phone in your hand. The main reason behind this is modern smartphone models and advanced AI features.

Why Is Mobile Video Editing Possible?

1. Powerful Smartphone Hardware

Modern smartphones come with:

- High-resolution cameras (up to 4K and 8K)
- Powerful CPUs and GPUs
- Sufficient RAM and storage

These make video cutting, color grading, and effects much easier.

2. Professional Mobile Video Editing Apps

There are now many mobile apps that work almost like desktop software, offering:

- Multi-layer timelines
- Transitions and visual effects

- Color correction and color grading
- Sound editing and background music

Using these apps, it's possible to create YouTube videos, short films, advertisements, and even cinematic trailers.

3. Easy to Use and Fast Workflow

The biggest advantages of mobile editing are:

- You can edit from anywhere
- Touchscreen controls make editing faster
- Easy to learn for beginners

That's why mobile editing is very popular among new content creators.

Limitations of Mobile Video Editing

Even though mobile editing is powerful, it has some limitations:

- Phones can heat up during large projects
- Very complex VFX work is difficult on mobile
- Long editing sessions can be tiring due to small screens

Still, for basic to mid-level video projects, mobile phones are more than sufficient.

With the addition of artificial intelligence, mobile video editing is becoming even easier and more powerful. Features like autocut, auto-color, and voice cleanup will turn mobile phones into even stronger video production tools in the future.

Popular Mobile Video Editing Apps

Today, there are many powerful mobile apps for video editing. Using these apps, you can create YouTube videos, Facebook reels, short films, and even advertisement-quality content.

1. CapCut

Cost:

- Free version: fully usable
- CapCut Pro: monthly or yearly subscription at low cost

Features:

- Multi-layer timeline
- Ready-made transitions and effects
- AI auto captions
- Background remover
- Color adjustment and filters
- Music and sound effects library
- 4K export

Best for:
Beginner and intermediate content creators, social media videos

2. KineMaster

Cost:

- Free (with watermark)
- Premium: monthly subscription

Features:

- Multi-layer video and audio
- Chroma key (green screen)
- Speed control
- Keyframe animation
- Professional audio tools
- Export up to 4K

Best for:
YouTubers, short film makers

3. VN Video Editor

Cost:

- Completely free
- No watermark

Features:

- Multi-track timeline
- Preset color grading

- Custom transitions
- Speed ramping
- PC-like interface
- 4K export

Best for:
Students, budget creators, serious editors

4. LumaFusion (iOS)

Cost:

- One-time purchase for lifetime use
- Some add-ons sold separately

Features:

- 6+ video tracks
- Professional color grading
- LUT support
- Advanced audio mixing
- Magnetic timeline
- Cinematic export

Best for:
Professional video editors, filmmakers

5. Adobe Premiere Rush

Cost:

- Free (limited exports)
- Paid subscription

Features:

- Simple timeline
- Built-in templates
- Auto color and auto sound
- Cloud sync (mobile and PC)
- Social media optimized export

Best for:
Beginners and Adobe users

6. InShot

Cost:

- Free (with watermark)
- Premium subscription

Features:

- Easy cut and trim
- Reels and TikTok formats
- Filters and effects

- Music and text tools
- HD export

Best for:
Reels, short videos, social media users

COMPARATIVE SUMMARY (I

APP	FREE USE	WATERMARK
CapCut	✓	✗
KineMaster	✓	✓
VN	✓	✗
LumaFusion	✗	✗
Premiere Rush	✓	✗
InShot	✓	✓

Among all the apps on this list, my personal favorite is **CapCut**, and I've edited every video I've made so far using CapCut.

That's why I'll now try to teach you all the professional editing details of CapCut step by step, with visuals.

Using CapCut may cost you around **3,000 to 5,000 taka per year**, but the good news is that you can also edit videos for free. For that reason, Google and YouTube have released a video editing app called **YouTube Create**.

So in this book, I will focus mainly on these two video editing apps, so that you can learn to edit your videos on your own and become a successful content creator.

What Is YouTube Create?

YouTube Create is the official mobile video editing app from Google/YouTube. It is mainly designed for creating, editing, and directly uploading videos to YouTube.
It is still relatively new, but new features are being added gradually.

Cost

- Completely free
- No watermark
- No paid subscription (at the moment)

This is a big advantage for new creators.

Main Features and Functions

1. Basic Video Editing

- Cut, trim, split
- Rearrange clips
- Multi-clip timeline

2. Music and Sound

- Free music library (copyright-safe)
- Add background music
- Audio volume control
- Voice-over recording

3. Text and Subtitles

- Add text
- Multiple font styles
- Auto captions (available in some regions)
- Subtitle timing adjustment

4. Effects and Filters

- Basic transitions
- Light color filters
- Blur and simple visual effects

5. Stickers and Graphics

- Emoji and graphic stickers
- YouTube-friendly visual elements

6. Direct YouTube Upload

- Upload directly to YouTube from the app
- Options to write title and description
- Supports both Shorts and long videos

What Type of Videos Is It Best For?

YouTube Create works best for:

- Vlogs
- Talking-head videos
- YouTube Shorts
- Educational videos
- Simple cinematic edits

Limitations

Despite all its features, there are some limitations:

- No advanced color grading
- No LUT support
- No green screen (chroma key)
- Heavy cinematic effects are not possible
- No iOS version yet (Android-only in many regions)

For professional-level work, other apps may be required.

YouTube Create vs Other Apps

Subject	YouTube Create	CapCut
Free	✓	✓
Watermark	✗	✗
Cinematic Editing	✗	✓
Beginner Friendly	5/5 ★★★★★	4/5 ★★★★☆
YouTube Focused	5/5 ★★★★★	3/5

Video Editing with the CapCut App

Video Editing with the CapCut App

In today's digital age, video content is a powerful communication tool. Whether it's social media, YouTube, online education, or personal memories, the need for video editing is growing everywhere. To meet this need, **CapCut** is a simple and effective mobile app. In the past, editing was done only on computers. Now, each smartphone has become like a computer, and CapCut is leading the way.

Introduction to the CapCut App

CapCut is a popular and modern video editing app designed mainly for smartphone users. It is available on both Android and iOS platforms. Using this app, you can easily cut and trim videos, add text, include music, apply filters, and use various visual effects. Especially for beginners, CapCut is a simple and user-friendly app.

Advantages of Editing Videos with CapCut

1. Easy-to-Use Interface

CapCut has a clean and simple interface, which allows new users to learn video editing in a very short time.

2. Advanced Features for Free

Many premium-quality features are available completely free in CapCut, which usually require payment in other apps.

3. Wide Range of Effects and Filters

CapCut offers many video effects, transitions, and filters that make videos more attractive and professional.

4. Text and Subtitle Support

You can easily add text, captions, and subtitles to videos. There are also options to change fonts, colors, and animations.

5. Music and Sound Effects

Users can add their own music, as well as use music and sound effects from CapCut's built-in sound library.

6. High-Quality Video Export

Edited videos can be exported in HD and Full HD resolution, which is suitable for sharing on social media platforms.

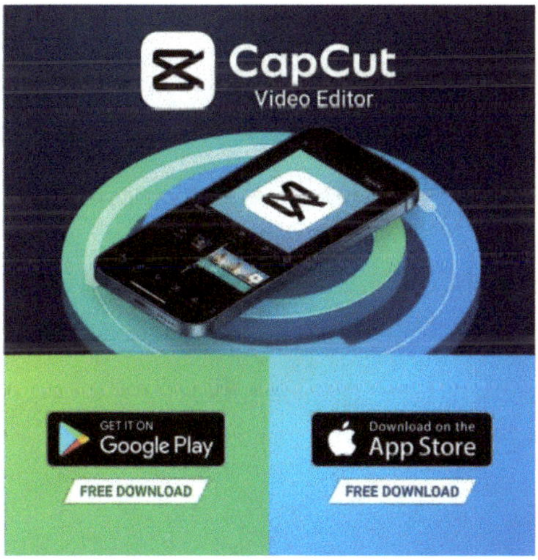

"Android users can find this app on the Play Store, and iPhone users can find it on the App Store. Once the app is downloaded, you'll see an interface like the one shown below

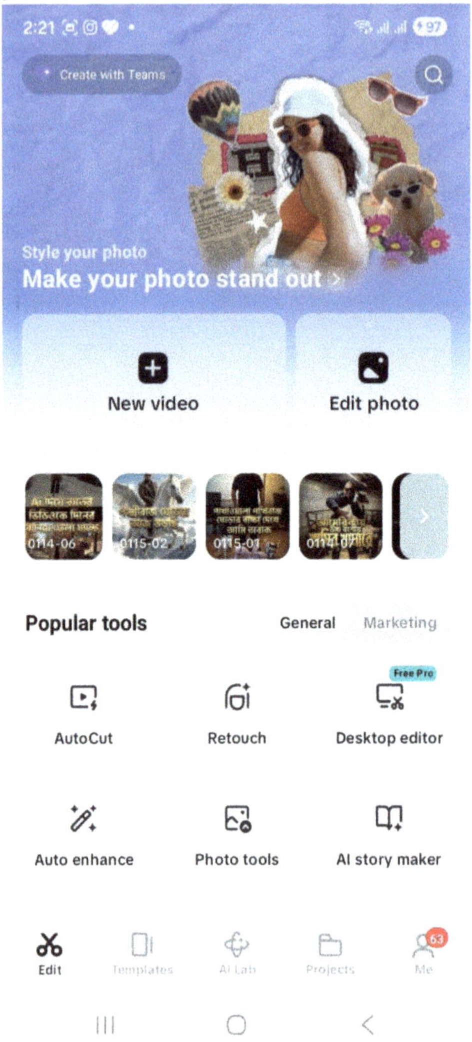

You will mainly use this app for video and photo editing. With this app, you can easily create professional-quality work in less time.

You can:

- Create new videos
- Edit any photo
- Use AI for automatic editing

The app is designed so that you can start working from the very first day.

When you open the app, the first screen you see is the **Home Screen**.
All your work will start from this screen.

Here, you will see several main options.

New Video

If you want to create a new video, click on the **New Video** option.

Here:

- You can select photos or videos from your phone's gallery
- Create a new video using the selected files

You will use this option when:

- You want to create Reels
- You want to make YouTube Shorts
- You want to create memory or highlight videos

This will be the first step of your editing process.

Edit Photo

If you want to enhance or edit a photo, click on the **Edit Photo** option.

Here:

- You select a photo
- Crop the image
- Adjust lighting, colors, and filters

You will use this option:

- Before posting on social media
- To make photos cleaner and more attractive

Project List :

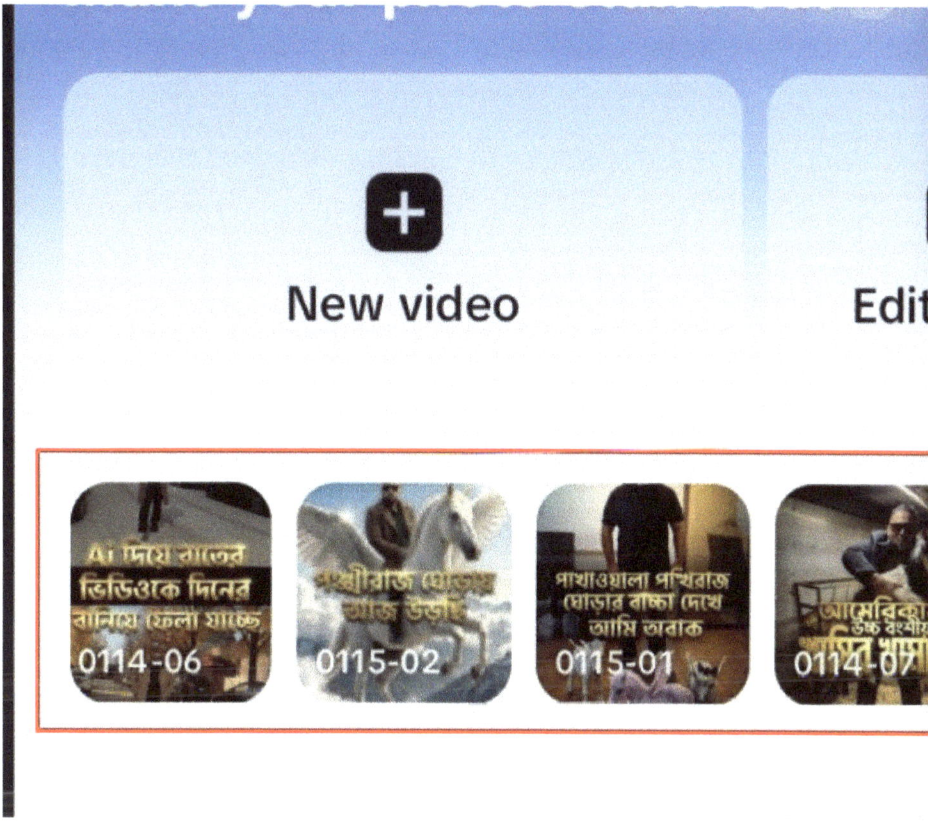

Below the **New Video** and **Edit Photo** options, you'll see some small thumbnails.
These show:

- Videos or photos you edited earlier
- The time of each project is displayed next to it

If you want to open any previous project again:

- Tap on that thumbnail
- Continue editing from where you left off

Popular

In this section, you'll find the most useful tools in the app.

AutoCut

If you want AI to automatically cut your video:

- Tap on the **AutoCut** option
- Select a video
- Let AI handle the rest

This saves a lot of time.

Retouch

If you want to improve face or skin appearance:

- Use the **Retouch** option
- Smooth skin and reduce blemishes

This works for both photos and videos.

Desktop Editor

If you want to work on a bigger screen:

- Use the **Desktop Editor** option
- Do professional-level editing

However, since this book focuses only on **mobile editing**, detailed instructions are given only for mobile.
If you purchase a CapCut subscription, you can also use it on a computer for premium work, but the computer version with subscription costs more.

Auto Enhance

If you want to improve photos or videos with one click:

- Tap on **Auto Enhance**
- Brightness, color, and contrast will be adjusted automatically

Photo Tools

For photo editing:

- Use **Photo Tools**
- Crop images, change backgrounds, and apply filters

AI Story Maker

If you want AI to automatically create videos or stories:

- Use the **AI Story Maker** option
- Select photos
- AI will arrange everything into a video

Editing Video with CapCut

To start editing a video in CapCut, click on **New Video**. Then bring the photos or videos you want to edit from your mobile into the CapCut app.

Here, if you select **Photos** (marked in red), you'll see only photos.
If you select **Video**, you'll see only the videos stored on your phone.

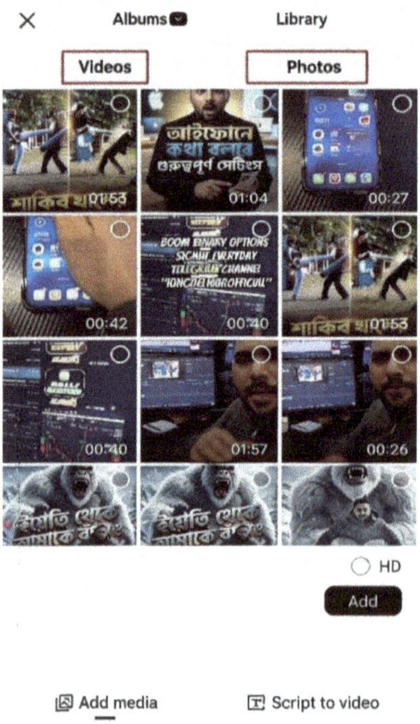

I'm going to edit a video, so I selected the video option

Once you click the 'Add' button, you'll be taken to the video editing interface.

Once you click the 'Add' button, you'll be taken to the video editing interface.

Left-Side Options

🔇 Mute Clip

- Turns off the original sound of the video

✂️ AI Clipper

- Automatically cuts the video using AI

- Removes unnecessary parts on its own

🖼 **Cover**

- Set the video thumbnail or cover image

Right-Side Layer Options

🎵 **Add Audio**

- Add music, voice, or sound effects

🅰 **Add Text**

- Add text
- Customize font, color, and animation

Bottom Menu Bar (Most Important)

✂️ **Edit**

- Cut, trim, split
- Speed, reverse, crop, rotate

🎵 **Audio**

- Add music, record voice
- Sound editing and volume control

🅰 **Text**

- Text and captions

- Font, animation, and style

✨ **Effects**

- Video effects (Glitch, Blur, Cinematic, etc.)

🖼 **Overlay**

- Place another video or image as a layer
- Create picture-in-picture

📝 **Captions**

- Auto subtitles
- Converts speech to text (Bangla/English)

If you slide the options to the right, you'll find even more editing tools.

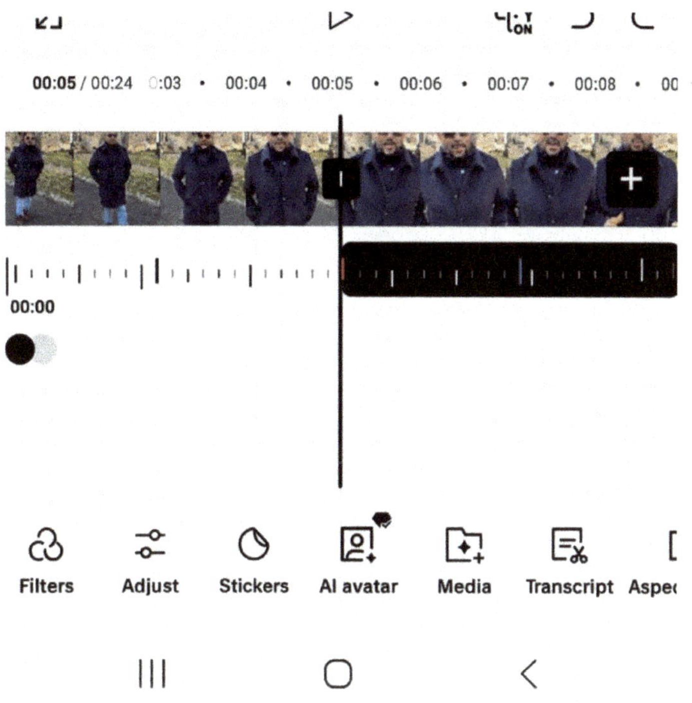

Filters

- Change the video's color and mood with one tap
- Gives cinematic, warm, cool, or vintage looks
- Very useful for reels and vlogs
- You can increase or decrease intensity

Adjust

- The section for manual video correction
- Brightness (light)

- Contrast (difference between dark and light)
- Saturation (color depth)
- Sharpness, exposure, highlights, shadows
- This is where professional color grading is done

Stickers

- Add emojis, GIFs, and animated stickers
- Mostly used for fun reels or casual videos
- You can control size, position, and timing

AI Avatar

- Creates a talking avatar using AI
- Useful when you don’t want to show your face
- Commonly used in educational or explainer videos

Media

- Add new media from your phone:
    - Video
    - Photo
    - Screenshot
- Very important for creating overlays

Transcript

- Converts spoken words into text (speech to text)
- Used to create auto subtitles
- Supports both Bangla and English (depending on the video)

Aspect

- Set the video size and ratio
- YouTube → 16:9
- Reels / TikTok → 9:16
- Square post → 1:1
- Always choose according to the upload platform

Background

What is Background?

- The area behind the video
- Used to fill empty space when making vertical or square videos

Where to find Background options?

- Inside Canvas / Background / Aspect

- Sometimes under Edit → Background (depends on CapCut version)

What can you do with Background?

- Solid color backgrounds (black, white, gradient)
- Blur background (blurred copy of the same video)
- Image or video backgrounds
- Texture or pattern backgrounds

When should you use it?

- Turning horizontal videos into reels
- When you want a cinematic blur
- When text needs more focus

Tips

- Cinematic look → Blur background + slight dark
- Talking videos → Solid dark background
- Fashion or lifestyle → Soft gradient

What Is a Template?

- A ready-made video design
- The entire video gets edited with one click

Where are templates found?

- In the CapCut app's Home or Template tab

- Outside the main edit screen

What can you do with templates?

- Auto cut and beat sync
- Automatic text animation
- Automatic transitions
- Music added automatically

When should you use templates?

- When you're short on time
- To follow Reels or TikTok trends
- When you want a professional look with minimal editing skills

Steps to Use a Template

1. Go to the Template tab
2. Choose a template you like
3. Add your own video or photos
4. CapCut edits it automatically
5. Export the video

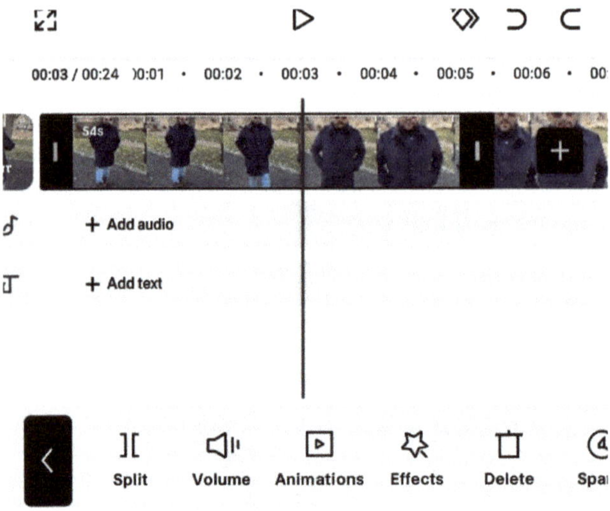

Now look at the image above. Since I clicked on the video, you'll see a black border around it. On your mobile, this border will appear white. For this book, the background has been changed to make it easier to read.

Now some new options have appeared. Let's get familiar with these options, because it will make the editing tutorial easier to follow.

Split

- Splits a single clip into two parts
- Useful for removing unnecessary sections

Volume

- Increase or decrease the clip's sound

- You can also mute it

Animations

- Adds entry and exit animations to the clip
- Includes Fade In, Fade Out, Slide, Zoom, and more

Effects

- Adds visual effects
- Creates cinematic, blur, or glitch-style looks

If you slide these options a bit to the right, you'll see even more options like in the image below.

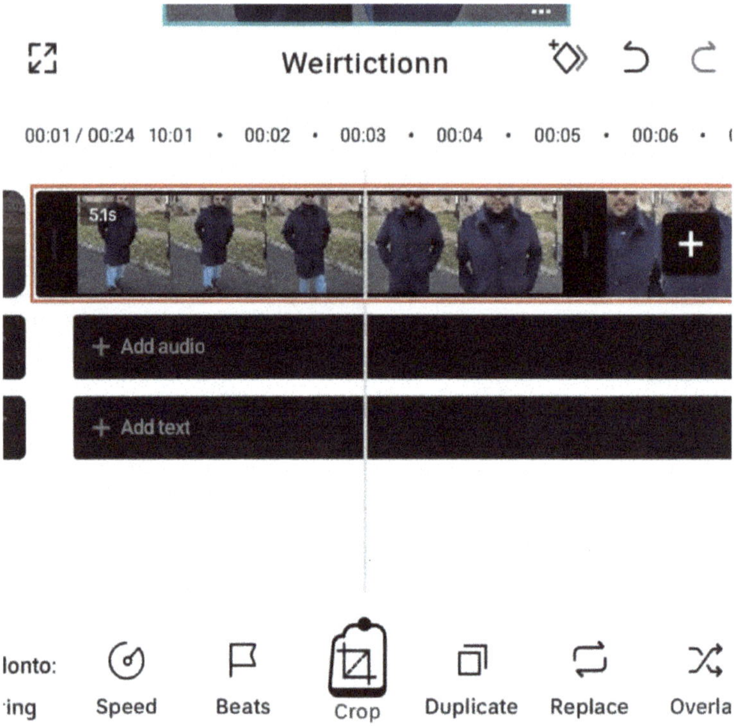

Speed

- Increase or decrease the video speed
- Used to create slow motion or fast motion
- You can change the speed of specific parts separately

Use cases:

- Dramatic walk → Slow
- Transitions or beat matching → Fast

Beats

- Cuts video according to the music beat
- Automatically detects beats and places markers
- Very useful for Reels and TikTok

How to use:

- Add music → Beats → Auto
- Sync clips to the beat

Crop

- Crop and resize the video
- Remove unnecessary sides
- Focus on the face or main subject

Use cases:

- Face zoom
- Remove extra background

Duplicate

- Creates a copy of the same clip
- Useful for testing effects or overlays

Use cases:

- Before/after edits
- Double effect layers

Replace

- Replaces the current clip with a new one
- Keeps the timing, only changes the content

Use cases:

- Wrong clip selected
- Replace with a better take

Overlay

- Place one video or image on top of another
- Create picture-in-picture effects
- Build cinematic layers

Use cases:

- B-roll footage
- Light leaks or texture overlays

Delete

- Completely removes the selected clip

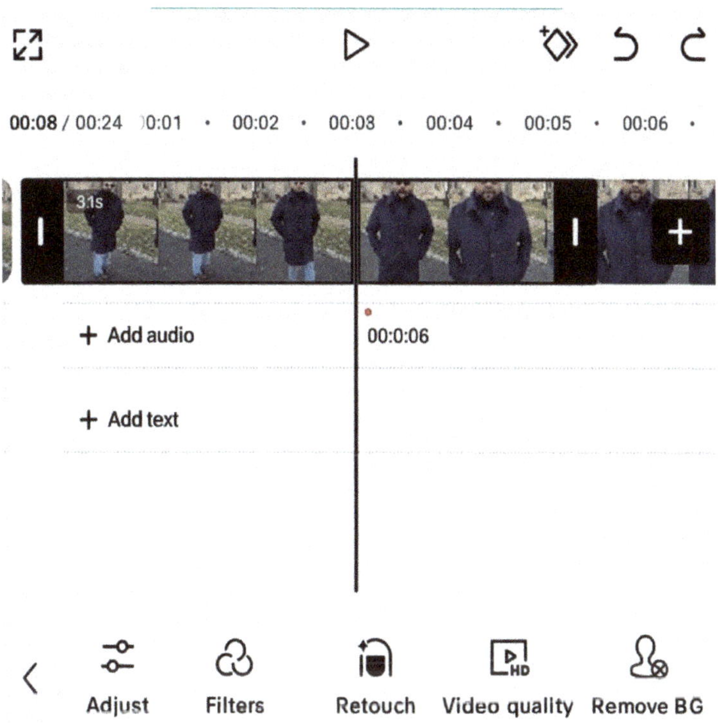

Adjust

Manually control the video's lighting, color, and details:

- **Brightness** → Increase or decrease light
- **Contrast** → Difference between dark and light areas
- **Saturation** → How strong the colors are
- **Exposure** → Overall light control
- **Highlights** → Reduce overly bright areas
- **Shadows** → Brighten dark areas
- **Sharpness** → How clear the video looks

Use cases:

- Fix raw footage
- Cinematic color grading

Filters

Change the overall look of the video with one tap:

- Ready-made color styles
- Warm, cool, vintage, cinematic looks
- Filter intensity can be increased or decreased

Use cases:

- Quickly enhance reels
- Maintain a consistent visual style

Retouch

Tools to enhance face and skin:

- Smooth skin
- Brighten the face
- Reduce dark circles or blemishes
- Slightly adjust face shape

Use cases:

- Talking videos

- Vlogs or portrait clips

Video Quality (HD / AI)

Enhance video quality using AI:

- Sharpens blurry or low-quality videos
- Improves detail and clarity
- Makes old footage look newer

Use cases:

- Improve mobile-shot videos
- Get a sharper look for reels

Remove BG

Remove the background from a video:

- Separates the subject
- Works without a green screen
- Allows you to add a new background

Use cases:

- Profile videos
- Creative background changes

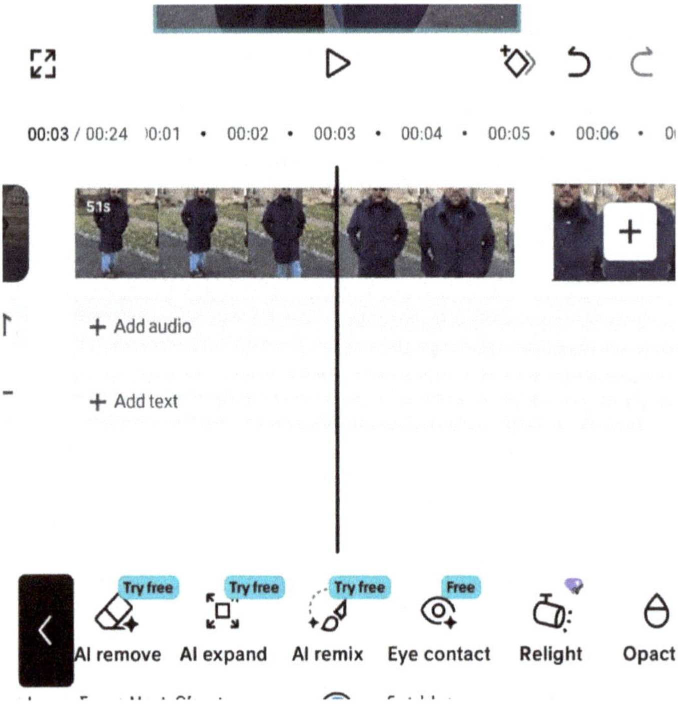

How to Use AI Remove

If you notice something unnecessary in your video, such as:

- A person in the background
- Unwanted objects on the road
- Something accidentally appearing in the frame

That's when you should use **AI Remove**.

What to do:

- Select the video clip on the timeline

- Tap the **AI Remove** option from the bottom
- Mark the object you want to remove using your finger or brush
- Confirm the action

After that:

- AI will automatically remove that part from the video
- It will intelligently fill the empty space by matching the background

Use this tool whenever you want a clean and focused video.

How to Use AI Expand

If you see that:

- The video was shot too close
- The frame feels too tight
- The aspect ratio doesn't match social media requirements

Then you should use **AI Expand**.

What to do:

- Select the video clip
- Tap on **AI Expand**
- Choose the size or direction you want to expand

After that:

- AI will automatically generate the surrounding areas of the video
- The frame will become larger without damaging the main footage

Use this tool:

- When creating reels or shorts
- When converting old videos into new formats

How to Use AI Remix

If you want:

- A more cinematic look
- A different color and mood
- Multiple styles from the same video

Then use **AI Remix**.

What to do:

- Select the video clip
- Tap on **AI Remix**
- Choose one style from the available options

After that:

- AI will change the color, contrast, and overall look
- Your video will get a completely new feel

Use this tool:

- For travel videos
- For fashion or lifestyle content

How to Use Eye Contact

If you make talking videos and notice:

- Your eyes are not aligned with the camera
- Your gaze keeps moving away

Then use **Eye Contact**.

What to do:

- Select the video clip
- Tap on the **Eye Contact** option
- Wait for the AI processing to finish

After that:

- AI will adjust your eye direction to face the camera
- The video will look more professional

Use this tool:

- For talking-head videos
- For tutorials or presentation videos

How to Use Relight

If you notice:

- The video is too dark
- The face is not clearly visible
- The lighting is poor

Then use **Relight**.

What to do:

- Select the video clip
- Tap on **Relight**
- Adjust how much light you want to add or reduce

After that:

- AI will fix the lighting automatically
- The video will look brighter and clearer

Use this tool:

- For indoor videos
- For night-time videos
- For footage shot in low-light conditions

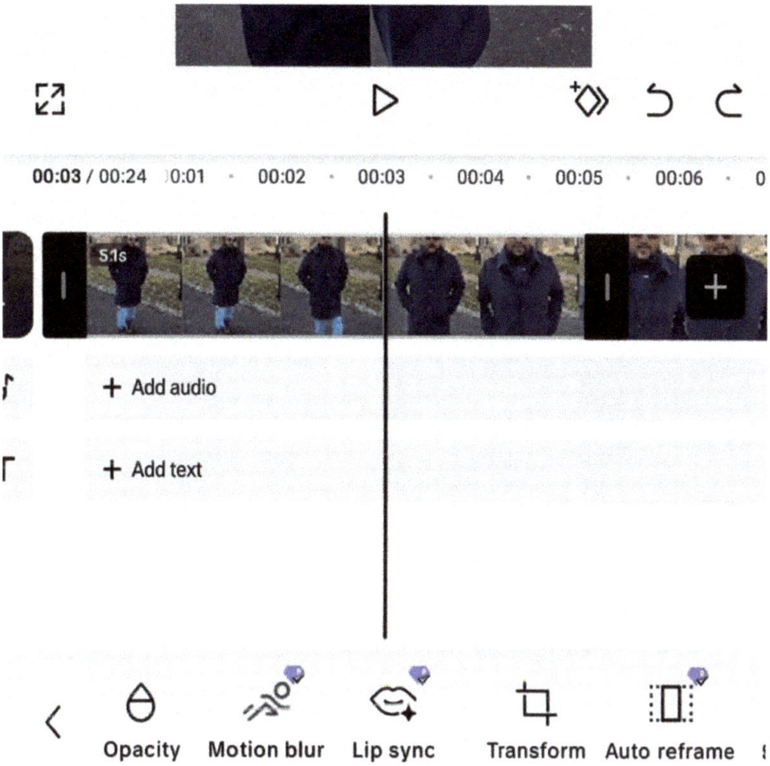

How to Use Opacity

If you want your video or any clip to:

- Look slightly transparent
- Show some of the background
- Give text or overlays a soft look

Then you should use **Opacity**.

What to do:

- First, select the video clip or text

- Tap on the **Opacity** option
- Adjust the slider to increase or decrease transparency

After that:

- Lower opacity makes the video look lighter and faded
- Higher opacity makes the video look stronger and more solid

Use this tool:

- To create a double-exposure–style look
- To soften the background

How to Use Motion Blur

If you notice:

- A lot of movement in the video
- Walking or running shots look too sharp or harsh

Then use **Motion Blur**.

What to do:

- Select the video clip
- Tap on **Motion Blur**
- Adjust the blur intensity

After that:

- Moving parts will get a slight blur

- The video will look smoother and more cinematic

Use this tool:

- For walking, running, or car shots
- To make action shots feel more natural

How to Use Lip Sync

If you:

- Make talking videos
- Use dubbing or voice-over
- Notice the lips don't match the audio

Then use **Lip Sync**.

What to do:

- Select the video clip
- Tap on the **Lip Sync** option
- Wait for the AI process to finish

After that:

- AI will match lip movements with the audio
- Talking videos will feel more realistic

Use this tool:

- For reels
- For tutorials

- For spoken videos

How to Use Transform

If you want to:

- Rotate a video
- Make it smaller or larger
- Change its position in the frame

Then use **Transform**.

What to do:

- Select the video clip
- Tap on **Transform**
- Adjust rotate, scale, or position

After that:

- The video will fit where you want it
- The frame will look more balanced

Use this tool:

- To fix wrongly shot videos
- To match social media aspect ratios

How to Use Auto Reframe

If you want to:

- Convert a landscape video into portrait
- Keep a person centered in the frame

Then use **Auto Reframe**.

What to do:

- Select the video clip
- Tap on **Auto Reframe**
- Choose your preferred aspect ratio

After that:

- AI will automatically follow the subject
- The video will stay in focus at all times

Use this tool:

- When making reels
- For YouTube Shorts
- For TikTok videos

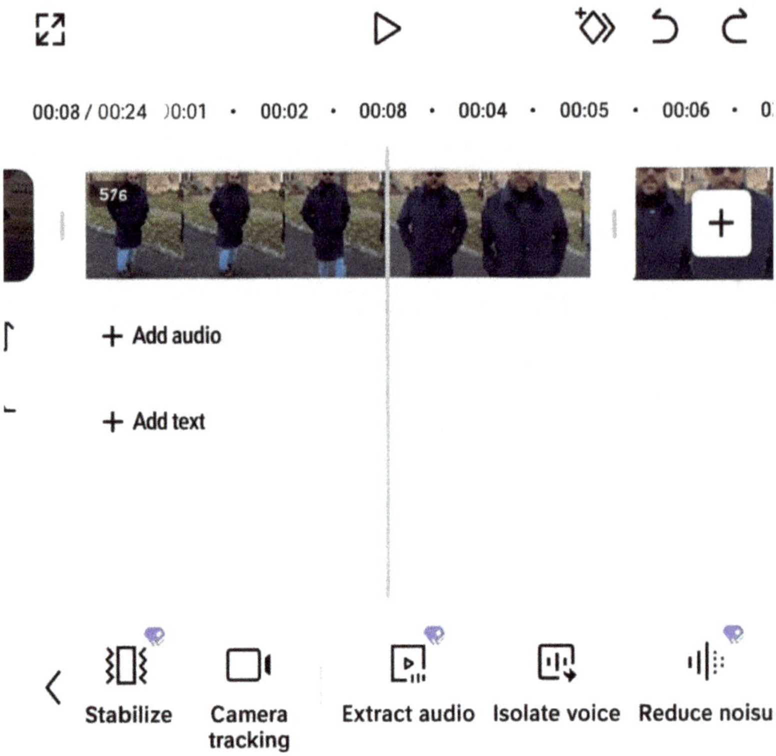

How to Use Stabilize

If you notice that:

- The video is shaky
- It was recorded while walking
- It was shot handheld

Then you should use **Stabilize**.

What to do:

- Select the video clip

- Tap on the **Stabilize** option from the bottom
- Wait until the processing is complete

After that:

- AI will reduce camera shake
- The video will look smoother and more comfortable to watch

Use this tool:

- For vlog videos
- For outdoor shots
- For handheld footage

How to Use Camera Tracking

If you want:

- A specific person or object to be followed in the video
- Text or effects to move along with that object

Then use **Camera Tracking**.

What to do:

- Select the video clip
- Tap on the **Camera Tracking** option
- Select the object you want to track

After that:

- AI will follow the movement of that object
- The video will look more dynamic

Use this tool:

- To make text follow a subject
- For creative video effects

How to Use Extract Audio

If you want:

- To separate only the audio from a video
- To reuse that audio in another video

Then use **Extract Audio**.

What to do:

- Select the video clip
- Tap on **Extract Audio**

After that:

- A separate audio track will be created from the video
- You can edit that audio independently

Use this tool:

- To separate voice
- To reuse music

How to Use Isolate Voice

If you notice that:

- Background noise is too loud
- Speech is not clear

Then use **Isolate Voice**.

What to do:

- Select the audio or video clip
- Tap on **Isolate Voice**

After that:

- AI will keep only your voice
- Background noise will be reduced

Use this tool:

- For talking videos
- For tutorials
- For vlogs

How to Use Reduce Noise

If you hear:

- Wind noise
- Car sounds
- Other distracting background noise

Then use **Reduce Noise**.

What to do:

- Select the audio clip
- Tap on the **Reduce Noise** option
- Adjust how much noise you want to remove

After that:

- Unwanted sounds will be reduced
- The audio will become clearer

Use this tool:

- For outdoor videos
- For old recordings
- For low-quality audio

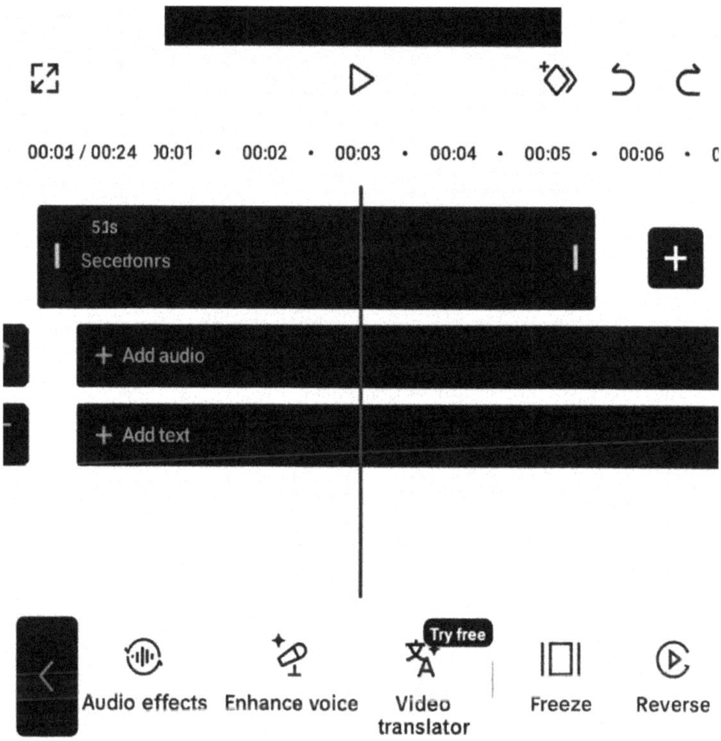

Inside **Audio Effects**, you'll find several more options, including **Voice Filters**, **Voice Characters**, and **Speech to Songs**.

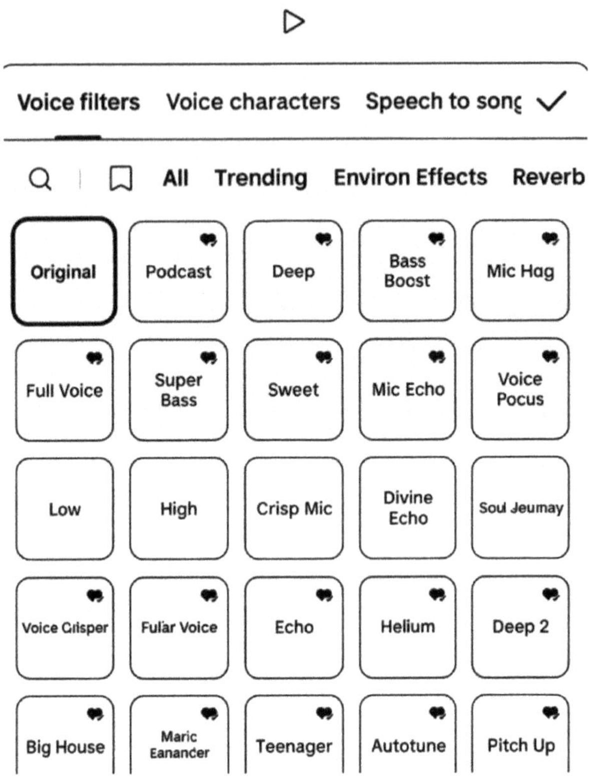

You'll use this screen to change the tone and style of your voice.
With AI, you can make your voice clearer, deeper, lighter, or more playful.

Original

Use this if:

- You don't want any changes
- You want to keep the original voice

Podcast / Crisp Mic / Voice Focus

Use these if:

- You want clearer speech
- You're making tutorials or talking videos

Deep / Deep 2 / Bass Boost / Super Bass

Use these if:

- You want a deep, powerful voice
- You want to add weight and strength to your voice

Sweet / Full Voice / Fuller Voice

Use these if:

- You want a soft and smooth voice
- You want a natural, pleasant sound

Echo / Mic Echo / Divine Echo / Big House

Use these if:

- You want echo or room effects
- You're making music or storytelling videos

Helium / Teenager / Pitch Up

Use these if:

- You want a fun or cartoon-style voice
- You're making humorous content

Autotune / Speech to Song

Use these if:

- You want to turn speech into music
- You want a musical-style voice

CapCut is being updated regularly, and you'll find the **Voice Characters** option right next to **Voice Filters**.

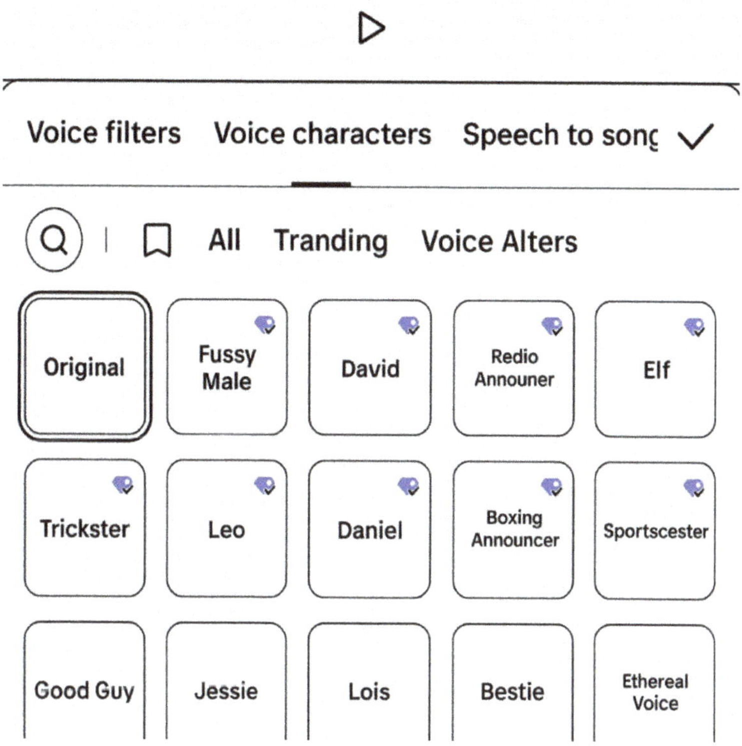

Voice Characters

You'll use this screen to make your voice sound like different characters.
Here, AI transforms your voice into various styles.

Original

Use this when:

- You want to keep your real voice
- You don't need any character-style voice

Fussy Male / David / Daniel / Leo

Use these when:

- You want different male voice styles
- You're making storytelling or narration videos

Radio Announcer / Sportscaster / Boxing Announcer

Use these when:

- You want a strong, announcement-style voice
- You're creating sports or news-style videos

Elf / Trickster

Use these when:

- You need fantasy or playful characters
- You're making cartoon or fun videos

Good Guy / Jessie / Lois / Bestie

Use these when:

- You want a friendly or soft voice
- You're making story-based or social content

Ethereal Voice

Use this when:

- You want a mysterious or dream-like voice
- You're creating cinematic or storytelling videos

If you want your video audio to:

- Sound more engaging
- Have a cinematic or creative feel
- Stand out from regular audio

Then you should use **Audio Effects**.

What to do:

- Select the video or audio clip
- Tap on **Audio Effects** from the bottom
- Choose one effect from the list

After that:

- Echo, room effects, or special touches will be added to the audio
- The mood of the video will change

Use this tool:

- For story videos
- For travel videos

- For creative content

How to Use Enhance Voice

If you notice that:

- Your voice doesn’t sound clear
- The voice feels weak or flat

Then use **Enhance Voice**.

What to do:

- Select the audio or video clip
- Tap on the **Enhance Voice** option
- Wait for the AI process to finish

After that:

- AI will clean up your voice
- Balance the volume
- Make speech sound clearer and sharper

Use this tool:

- For talking videos
- For tutorials
- For voice-over videos

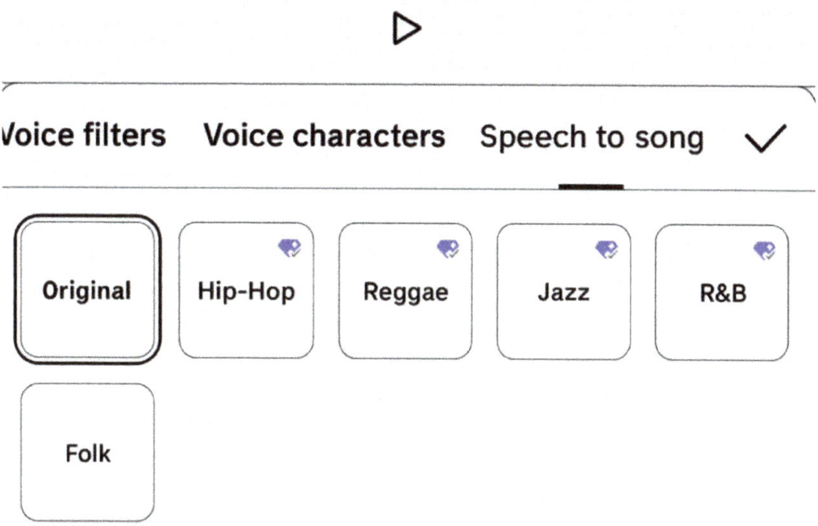

And the most fun part here is that you can turn **any voice into a song** if you want.

Hip-Hop

Use this if:

- You want a rap or beat-style song
- You need an energetic voice

Reggae

Use this if:

- You want a calm, relaxed music vibe
- You're making light and soft music videos

Jazz

Use this if:

- You want a smooth, classic music style
- You're creating slow or elegant videos

R&B

Use this if:

- You want a soft, romantic singing voice
- You're making moody or emotional content

Folk

Use this if:

- You want a simple, natural, or traditional music style
- You're making storytelling or cultural videos

That covers everything about **Audio Effects**.
Now, let's go back to our previous discussion.

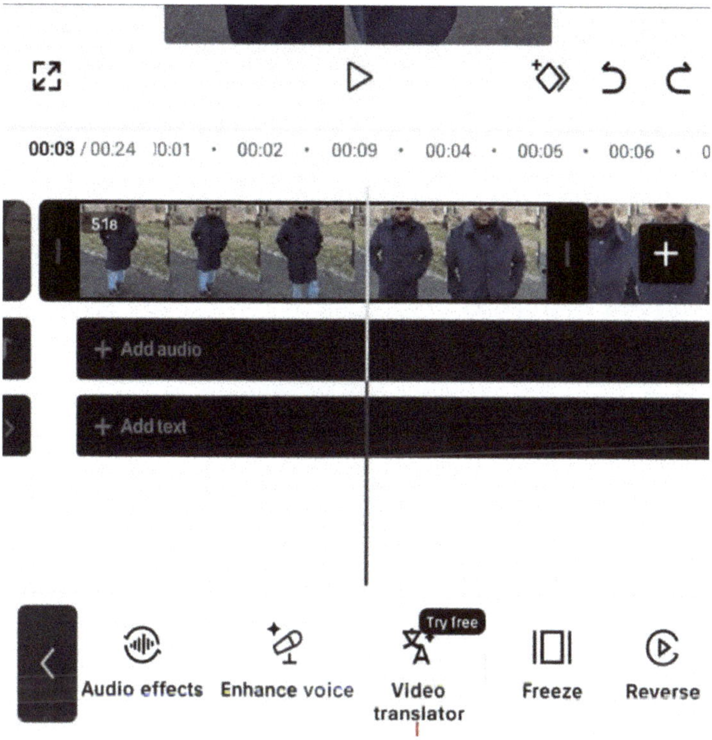

How to Use Video Translator

If you want to:

- Convert a video from one language to another
- Add voice or subtitles in a different language

Then you should use **Video Translator**.

What to do:

- Select the video clip
- Tap on the **Video Translator** option
- Choose the source language and the target language

After that:

- AI will understand the speech in the video
- It will add voice or text in the new language

Use this tool:

- To create videos for an international audience
- To repurpose one video into multiple languages

How to Use Freeze

If you want to:

- Pause a specific moment in the video
- Bring focus to a particular scene

Then use **Freeze**.

What to do:

- Move to the desired point on the timeline
- Select the video clip
- Tap on the **Freeze** option

After that:

- The video will stop on that frame
- The same frame will stay on screen for a short time

Use this tool:

- To highlight important moments

- To add text or explanations

How to Use Reverse

If you want to:

- Play the video backward
- Create fun or creative effects

Then use **Reverse**.

What to do:

- Select the video clip
- Tap on the **Reverse** option
- Wait for the AI process to finish

After that:

- The video will play from end to beginning
- It adds a creative and entertaining touch

Use this tool:

- For fun videos
- For transition effects
- For creative content

Now, if we **unselect the same video**, we'll see a few more options again. I'll explain how to use these options now.

Edit

- Cut, trim, and split videos
- Increase or decrease speed
- Crop, rotate, or reverse

All the basic tools needed to fix and shape the main video.

Audio

- Add music or background music
- Record voice-over
- Add sound effects
- Adjust volume, fade in, and fade out

After clicking on **Audio**, you'll find several more options. I'll explain those in detail next.

Extract

- Separates audio from a video
- Used to extract original sound or dialogue
- Helpful before replacing music

Sounds

- Add music from the CapCut library
- Includes trending tracks
- Organized by genre

Sound FX

- Add sound effects
- Whoosh, pop, click, transition sounds
- Makes the video more lively

Record

- Record your own voice
- Used for voice-overs
- Helpful for tutorials or explanation videos

AI Voiceover

- Creates AI-generated voice from text
- Multiple male and female voice options
- Good for clear narration

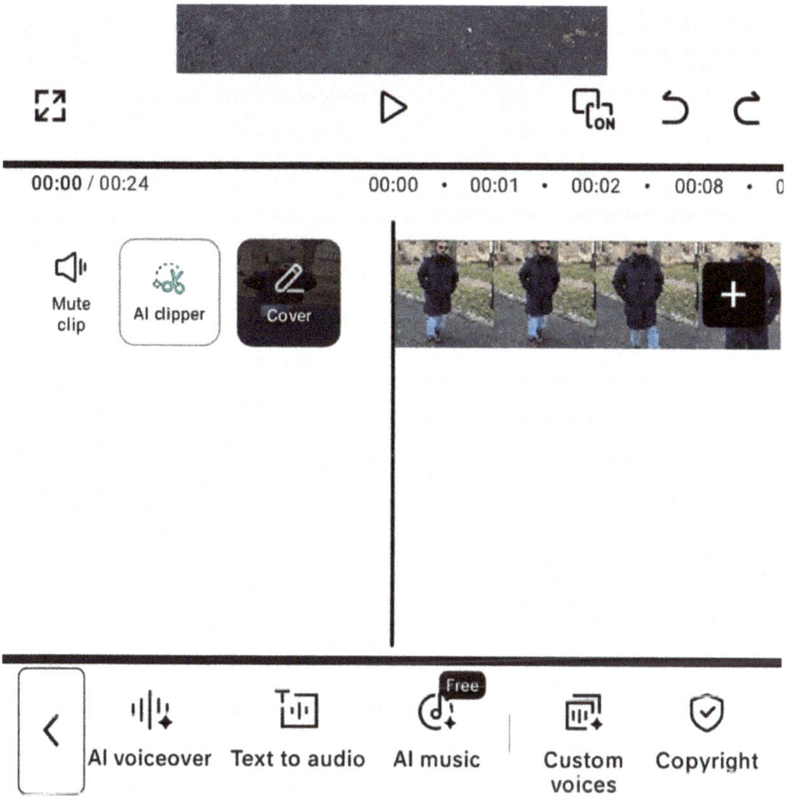

Text to Audio

- Converts written text into voice
- Voice tone and speed can be adjusted
- Allows quick voice-over creation

Use cases:

- Tutorials
- Short reel narration

AI Music

- Generates background music using AI
- Creates music based on mood
- Provides copyright-safe tracks

Use cases:

- Vlogs
- Cinematic reels

Custom Voices

- Uses your own voice as an AI voice
- The same voice can be generated repeatedly
- Great for branding and consistency

Use cases:

- Personal channels
- Regular content

Copyright

- Checks whether music is copyright-safe
- Helps avoid platform issues
- Ensures safe uploads

Use cases:

- YouTube
- Facebook
- Instagram

Text

- Add text
- Create subtitles or titles
- Change font, color, and size
- Apply text animations

Inside the **Text** section, you'll find several more options that help make your videos look more professional with different text styles.

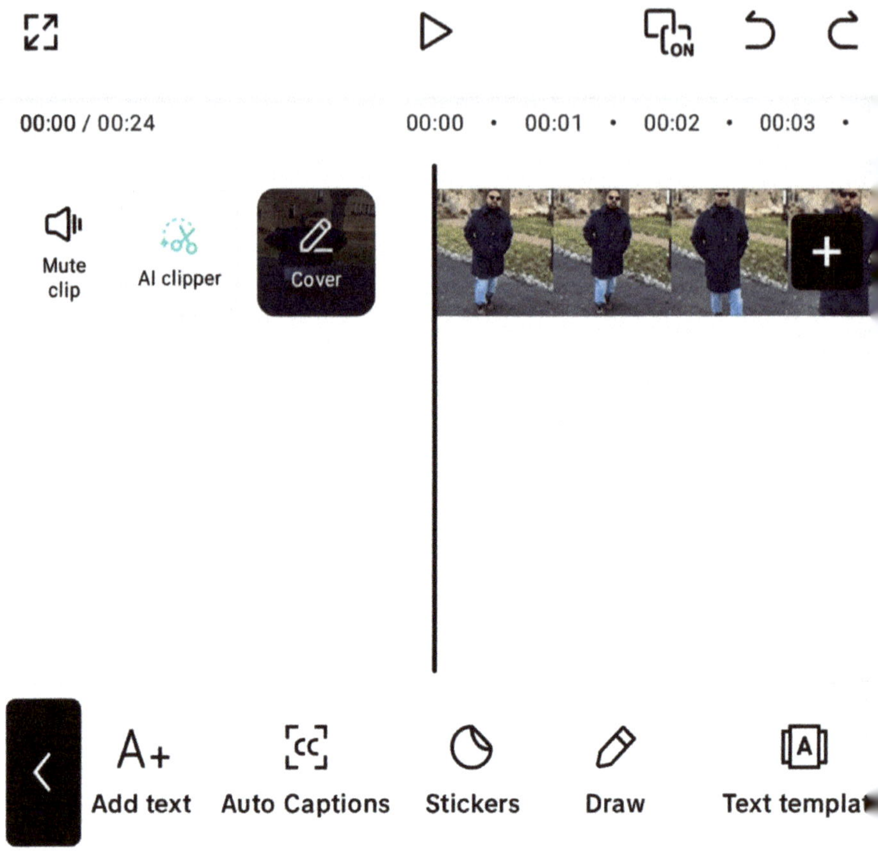

Add Text

- Add text manually
- Used for titles, subtitles, names, or quotes
- You can set font, color, size, and animations

Auto Captions

- Automatically converts spoken words into text
- Creates subtitles quickly
- Language and caption style can be changed

Stickers

- Add emojis, animated stickers, or GIFs
- Great for fun or casual videos
- You can control size and timing

Draw

- Draw or highlight directly on the screen
- Use arrows or circles to show focus
- Very useful for tutorial videos

Text Templates

- Ready-made text styles
- Animations and fonts are pre-set
- Helps you add stylish text quickly

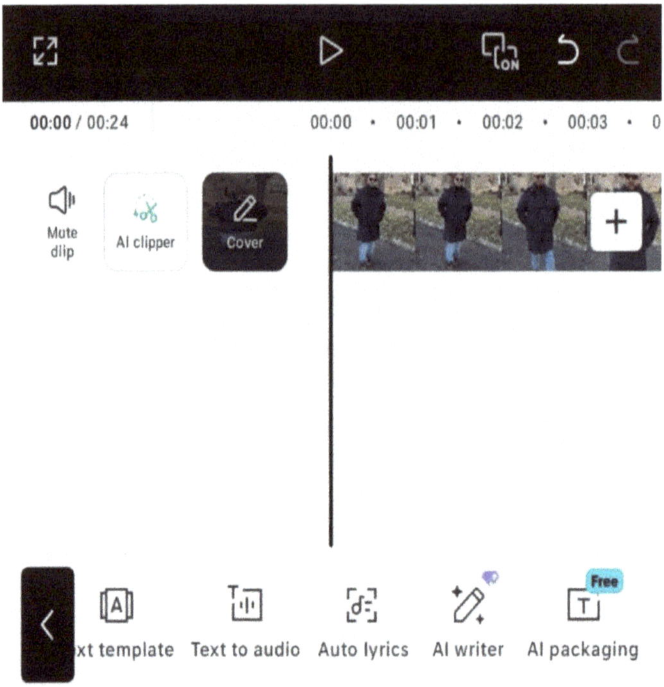

Inside **Text**, there are several more important features that you can see in the image above.

Text to Audio

- Converts written text into voice
- Uses AI-generated voices
- Helps create voice-overs very quickly

Used for:

- Explainer videos
- Faceless content

Auto Lyrics

- Automatically adds lyrics synced with the song
- Text matches the beat of the music
- Very useful for music videos or reels

Used for:

- Song reels
- Lyric videos

AI Writer

- Uses AI to generate captions or text
- Can create quotes, hook lines, or short scripts
- Saves a lot of time

Used for:

- Social media captions
- Intro text

AI Packaging

- Automatically styles the entire text using AI
- Sets font, animation, and placement on its own
- Gives a clean and professional look

Used for:

- Quick, polished text design
- Maintaining consistent branding

Effects

- Adds visual effects to the video
- Includes blur, cinematic, glitch, and light effects
- Changes the mood and style of the video
- Makes the video more eye-catching

However, inside **Effects**, there are a few more important options that we'll look at next.

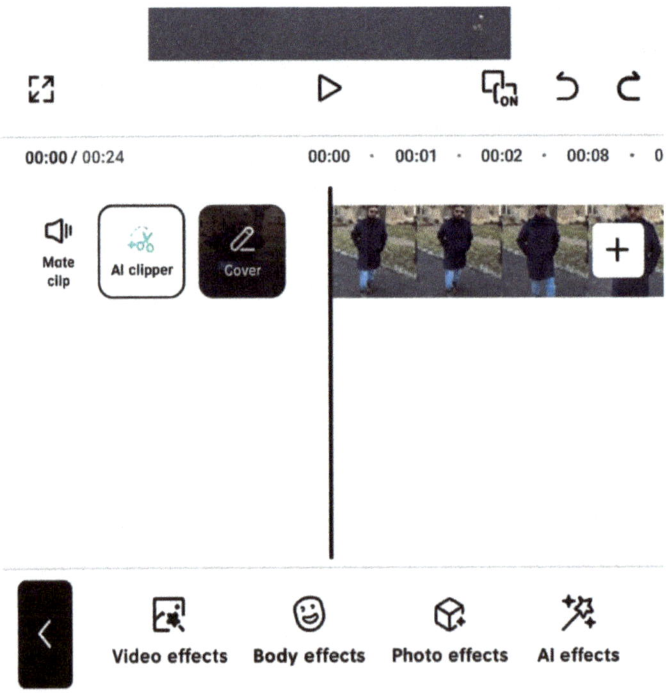

Video Effects

- Adds visual effects to the entire video
- Includes blur, cinematic, glitch, light, shake, and more
- Changes the mood and overall style of the video

Used for:

- Creating a cinematic look
- Making reels more eye-catching

Body Effects

- Applies effects to the human body or movement
- Glow, outline, motion trail, clone-type effects
- Keeps the subject as the main focus

Used for:

- Dance videos
- Creative reels

Photo Effects

- Adds photo-style effects
- Grain, vintage, blur, light textures
- Works best for static or slideshow-style videos

Used for:

- Photo reels
- Aesthetic edits

AI Effects

- Applies smart effects using AI
- Auto background changes, motion effects, glow, style shifts
- Delivers a professional look with very little effort

Used for:

- Trendy reels
- Fast editing workflows

Overlay

- Places one video or image on top of another
- Creates picture-in-picture effects
- Used to add logos, textures, or B-roll

Used for:

- Creative layering

Captions

- Displays spoken words as on-screen text

- Automatically generates subtitles
- Font, color, and style can be customized

Benefits:

- Improves understanding
- Increases reach and engagement

This covers all the essential concepts you need to understand **before starting video editing**.
Now, scan the **QR code below** to watch the direct video tutorial. Many new topics are explained there. Since we've discussed a lot here, it's normal to feel a bit overwhelmed. Watching the video will make everything much clearer.

Edit Videos for Free with the YouTube Create App

YouTube Create is YouTube's own mobile video editing app. It is specially designed for creating YouTube Shorts and regular YouTube videos.

The main purpose of this app is to help new and small creators easily create videos for free and quickly. However, many people are now using it because videos can be edited for free. As a result, its popularity has increased significantly.

When was YouTube Create released?

- September 2023
- First released on Android
- Later gradually expanded to more countries

Developed by
YouTube (owned by Google)

Why did YouTube Create become so popular?

Beginner-friendly
• New creators can easily use it
• No editing knowledge is required

Made specifically for Shorts
• Automatic support for vertical video (9:16)
• Shorts-focused timing, cuts, and music

Free and No Watermark
• No subscription required
• No watermark on videos
This is a big advantage for creators

Direct integration with YouTube
• Videos can be uploaded directly to YouTube from the app
• No need for separate export and upload steps

Built-in AI Features
• Auto captions
• Beat-sync cuts
• Simple transitions
• Smart text tools
AI makes the editing process easier

What can you do in the YouTube Create app?

Video Editing
• Cut, trim, split clips
• Add multiple clips
• Smooth transitions

Text & Captions
• Add subtitles
• Auto captions
• Text animation

Audio
- Background music
- Sound effects
- Beat-based sync

Effects & Filters
- Simple effects
- Color filters
- Clean, YouTube-friendly look

To use this app, you must first log in with the Gmail account on your mobile phone. After logging in, you will see an interface like this.

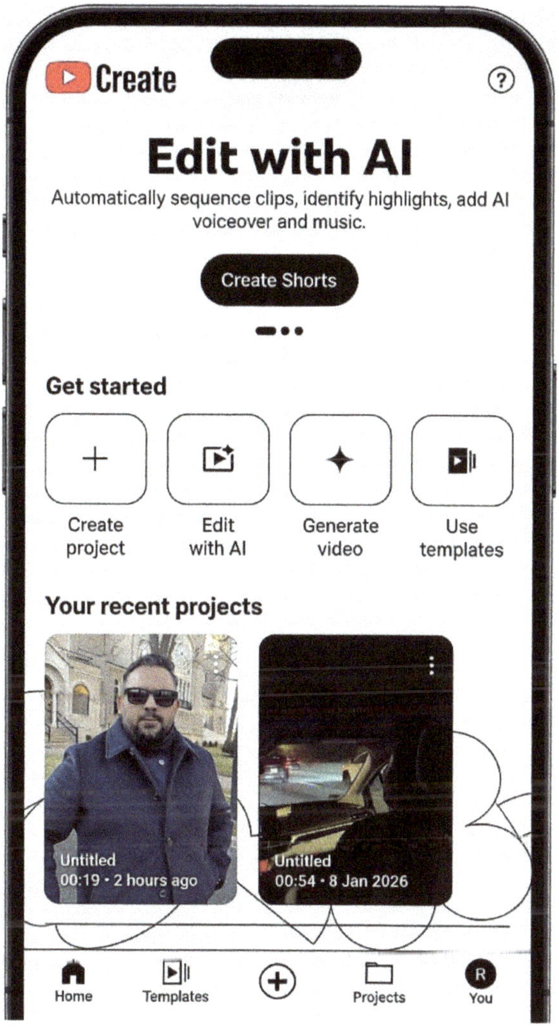

However, this app is still under development, so users may face some issues for now. These problems are expected to be resolved very soon. Being able to use a free app without a watermark can save a significant amount of money. After understanding the interface features, watching the direct video tutorial will make everything clear without any confusion.

Edit with AI

What it does

- Automatically arranges video clips
- Identifies important parts (highlights) using AI
- Removes unnecessary sections
- Can add AI voiceover and music

Why it is used

- For new creators
- To create videos in less time
- To work without having prior editing knowledge

Create Shorts

What it does

- Provides a separate AI mode for creating Shorts
- Sets the video to vertical (9:16) format
- Uses appropriate length and cuts for Shorts

Why it is used

- To create YouTube Shorts
- To convert long videos into Shorts

Create Project

What it does

- Creates a new blank project
- Allows users to manually edit videos
- Provides less AI assistance and more manual control

Why it is used

- For those who want to edit on their own
- To work simply but with personal customization

Edit with AI (Get Started section)

What it does

- AI handles the entire editing process
- AI selects clips, makes cuts, and chooses music

Why it is used

- To create videos very quickly
- For beginners or student creators

Generate Video

What it does

- Creates videos from text or ideas
- AI automatically arranges clips, music, and flow

Why it is used

- To create faceless content
- When you have an idea but no video footage

Use Templates

What it does

- Allows the use of ready-made video designs
- Animations and text styles are pre-set

Why it is used

- To create trendy Shorts
- To save time

Your Recent Projects

What it does

- Shows previously created or unfinished projects
- Allows you to resume previous work

Why it is used

- To maintain continuity
- To easily find old work

Bottom Navigation Bar

Home

- Main screen
- All AI options are available here

Templates

- View all templates

(Plus)

- Start a new project or video

Projects

- View all saved projects

You

- Profile and settings

Now, simply scan the QR code below to watch the complete video tutorial.

Instagram Edits

Capcut and Youtube Create Type another Free Video Editing APP is Instagram Edits.

ow, simply scan the QR code below to watch the complete video tutorial.

In Conclusion

In the present time, Artificial Intelligence (AI) is rapidly influencing almost every aspect of our personal and professional lives. From content creators to business owners, employees, and students—everyone now needs to leverage AI to present their work and skills on a global level. It is no longer enough to simply do good work; presenting that work effectively is equally important.

One of the most powerful tools for presentation is high-quality editing. Content creation, even when started from a small town, can now reach audiences across many countries around the world. Whether it is video, audio, or image-based digital content, editing significantly enhances its quality, acceptance, and impact. Many people still believe that good editing is only possible with a powerful or expensive computer. In reality, modern AI technology has completely changed this perception.

This book aims to provide a practical pathway out of that limited mindset. It demonstrates how professional-quality content can be created using AI-powered mobile apps and tools with just a regular smartphone. Through apps like CapCut and tools from YouTube such as YouTube Create, along with AI scripting, voiceovers, thumbnail design, image upscaling, and proper prompt usage, this book explains step by step how to make content more effective and engaging. However, reading the book alone is not enough—you must practice extensively, especially with prompts. Only then can you truly become an AI-Powered Content Creator.

The main objective of this book is to help readers understand that creating high-quality content does not require a large budget or powerful hardware. Instead, it requires the right knowledge, creative thinking, and intelligent use of AI technology. For those who want to make their work known on a larger scale, this book is designed to serve as a practical guide—and every chapter has been structured with that belief in mind.

www.ingramcontent.com/pod-product-compliance
Lightning Source LLC
LaVergne TN
LVHW050913210926
848482LV00012B/422